Complex Analyses in Engineering, Science and Technology

Authored By

S. G. Ahmed

Taif University
Al-Haweiah
Faculty of Science
Department of Mathematics and Statistics
Kingdom of Saudi Arabia

DEDICATION

To my professors around the world I learnt so much from, and my loving family who unconditionally supported me throughout my busy working life.

CONTENTS

Preface *i*

CHAPTERS

1. **Complex Numbers, Variables and Functions** 3
2. **Complex Functions** 29
3. **Methods of Evaluating Complex Integrals** 61
4. **Complex Series** 103
5. **Theory of Residue** 117
6. **Transformation and Conformal Mapping** 149

References 195

Subject Index 197

PREFACE

The present book is specifically designated for undergraduate students in mathematics, physics and engineering. Basic sciences especially mathematics does not hold much prominence amongst sciences in the Middle East countries. So we have designated and intended this book for the Middle East countries and others. Engineering students do not take into consideration proofs when dealing with any branch of mathematics, the situation is completely different for science students. Engineering students mainly need to know how to apply a theorem but science students are usually interested in how the theorem comes from. The book underhand can be considered as an introductory course in complex numbers, complex functions, transformations and others, all of which are the basis of general complex analysis. An important note for the reader is to know that when dealing with subjects such as complex-valued functions, they should be familiar with the functions in the real plane and also with so many topics in complex analysis have similarity in the real field, so we advise the reader to make a short review for functions in real plane, differentiation and integration rules before dealing with the corresponding subjects in the complex analysis. This present book consists of six main chapters starting with the basis of complex numbers, its ideas, history and other deeper subjects. We tried to put so many solved examples to help the students to very well understand all areas in the contents of the present book. We still intending to increase the contents of the book by adding in the nearest future, more chapters in a brief detail and more areas in the field of complex analysis such that all the present and future materials will be mainly for undergraduate engineering students.

ACKNOWLEDGEMENTS

Declared None.

CONFLICT OF INTEREST

The author confirms that this ebook contents have no conflict of interest.

S. G. Ahmed
Taif University
Al-Haweiah
Faculty of Science
Department of Mathematics and Statistics
Kingdom of Saudi Arabia

CHAPTER 1

Complex Numbers, Variables and Functions

Abstract: The complex numbers and the complex functions are main topics to go through the complex analysis. The complex analysis is a wide topic found in many areas of pure, applied and engineering applications. Therefore, many authors have tried to put their experience in this field that ignored by others. The present chapter aims mainly to cover some basic concepts for the beginners in the field of complex analysis, such as concept of complex number, idea of complex number, algebraic operations and more.

Keywords: Complex number, polar representation of complex number, complex number properties, roots of complex number, Cauchy theorem, generalized Cauchy theorem.

1.1. IDEA OF COMPLEX NUMBER

The complex number is represented mainly by a point in the complex plane. When the real part in the complex number equals zero, then it takes the name "Pure imaginary complex number" and when its imaginary part equals zero, then it takes different name called "Pure real complex number" [1].

1.2. HERITAGE OF COMPLEX NUMBER

To start this heritage, let us see how the effort done by Italian Mathematician Girolamo Cardano for solving the following cubic equation [2]:

$$z^3 + a_2 z^2 + a_1 z + a_o = 0 \tag{1}$$

Girolamo Cardano assumed the following:

$$z = x - \frac{a_2}{3} \tag{2}$$

Equation (1) which was then reduced to what was called "*Depressed Equation*" and so equation (1) took the following form named as "*Depressed Equation*":

$$x^3 + bx + c = 0 \tag{3}$$

In which:

$$b = a_1 - \frac{1}{3} a_2^2 \tag{4}$$

S. G. Ahmed

$$c = -\frac{1}{3} a_1 a_2 + \frac{2}{27} a_2^3 + a_o \tag{5}$$

One solution for the *Depressed Equation* is given as:

$$x = \sqrt[3]{\left(-\frac{c}{2} + \sqrt{\frac{c^2}{4} + \frac{b^3}{27}} \right)} + \sqrt[3]{\left(-\frac{c}{2} - \sqrt{\frac{c^2}{4} + \frac{b^3}{27}} \right)} \tag{6}$$

Before continuing, let us address an example to understand how the depressed equation can be obtained.

Example

Reduce the following cubic equation to depressed one $z^3 + 9z^2 + 24z + 20 = 0$

Solution

First of all, let us compare the given equation by the standard one, given by equation (1), one can get:

$$a_o = 20$$
$$a_1 = 24$$
$$a_2 = 9$$

By making use of the following substitution, $z = x - \frac{a_2}{3} = x - \frac{9}{3} = x - 3$ into the given equation, leads to:

$$(x-3)^3 + 9(x-3)^2 + 24(x-3) + 20 = 0$$

Simplifying, this results in:

$$x^3 - 3x + 2 = 0$$

Again, to find one root of the depressed equation, one can make use of relation given by equation (6):

$$x = \sqrt[3]{\left(-\frac{c}{2} + \sqrt{\frac{c^2}{4} + \frac{b^3}{27}}\right)} + \sqrt[3]{\left(-\frac{c}{2} - \sqrt{\frac{c^2}{4} + \frac{b^3}{27}}\right)}$$

According to the deduced depressed equation, $b = -3, c = 2$, the required root will take the following form:

$$x = \sqrt[3]{\left(-\frac{c}{2} + \sqrt{\frac{c^2}{4} + \frac{b^3}{27}}\right)} + \sqrt[3]{\left(-\frac{c}{2} - \sqrt{\frac{c^2}{4} + \frac{b^3}{27}}\right)} = \sqrt[3]{-1} + \sqrt[3]{-1} = -2$$

We have now one root of the equation then, the next step is to reduce the order of the equation using the synthetic division. This will make some quick revision to the subject of synthetic division through the following example.

Synthetic Division

The synthetic division is a method to divide two polynomials and the order of the numerator is greater than the denominator [3].

Example

Use synthetic division to divide $x^4 - 3x + 5$ by the factor $(x - 4)$

Solution

Step (1)

Write down the coefficients of the polynomial in a decreasing order and the term whose not exist its coefficient will be zero as shown below:

divisor x - c	coefficients of dividend				
4	1	0	0	-3	5

Step (2)

Pull down the coefficient of the largest power as shown:

4	1	0	0	-3	5
	↓				
	1				

Step (3)

Carry out the product and summation below

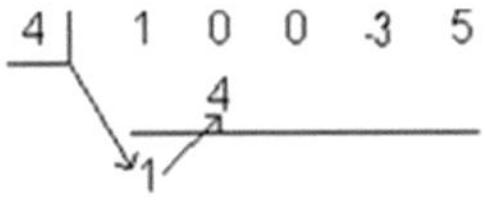

Step (4)

Write the sum in the bottom row

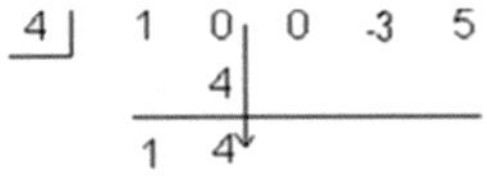

Step (5)

Repeat the above procedure as shown below

4	1	0	0	-3	5
		4	16	64	244
	1	4	16	61	249

Step (6)

Write down the final result as shown

$$x^3 + 4x^2 + 16x + 61 + \frac{249}{x-4}$$

The depressed equation had real solution, then two parts of x appears in equation (5), could be written as:

$$x = \sqrt[3]{\left(-\frac{c}{2} + \sqrt{\frac{c^2}{4} + \frac{b^3}{27}}\right)} + \sqrt[3]{\left(-\frac{c}{2} - \sqrt{\frac{c^2}{4} + \frac{b^3}{27}}\right)} \tag{7}$$

$$x = \left(u + v\sqrt{-1}\right) + \left(u - v\sqrt{-1}\right) \tag{8}$$

According to the present example, equations (7-8) can written in some more details as follows:

$$\left(u+v\sqrt{-1}\right)=\sqrt[3]{\left(-\frac{c}{2}+\sqrt{\frac{c^2}{4}+\frac{b^3}{27}}\right)}=\sqrt[3]{(2+11i)} \qquad (9)$$

$$\left(u-v\sqrt{-1}\right)=\sqrt[3]{\left(-\frac{c}{2}-\sqrt{\frac{c^2}{4}+\frac{b^3}{27}}\right)}=\sqrt[3]{(2-11i)} \qquad (10)$$

1.3. FORMS OF COMPLEX NUMBER

The complex number takes different forms as follows:

$$z=(x,y) \qquad (11)$$

Complex Conjugate Number

$$z=x+iy \qquad (12)$$

Idea of the Conjugate of the Complex Number

If one goes back to the equation $t^2=-1$, it will be noticed that it has two different roots, and the two roots are complex in nature. Based on this fact one can define the conjugate of the complex number as it is the same as the complex number but with different sign, as follows:

$$\bar{z}=x-iy \qquad (13)$$

Some Basic Rules

$$i=\sqrt{-1} \qquad (14)$$

$$i^2=\left(\sqrt{-1}\right)^2=-1 \qquad (15)$$

$$i^3=\left(\sqrt{-1}\right)^3=-i \qquad (16)$$

$$i^4=\left(\sqrt{-1}\right)^4=1 \qquad (17)$$

1.4. ALGEBRA OF COMPLEX NUMBERS

Let us have two complex numbers $z_1 = (x_1, y_1)$, $z = (x_2, y_2)$, respectively, then the successive algebraic operations will be shown in details below:

Addition and Subtraction

Both addition and subtraction of two complex numbers are defined as follows:

$$\begin{aligned} z_1 \pm z_2 &= (x_1 + iy_1) \pm (x_2 + iy_2) \\ &= (x_1 \pm x_2) + i(y_1 \pm y_2) \end{aligned} \tag{18}$$

Example

Add and subtract the following two complex numbers:

$$z_1 = (5 + 3i)$$

$$z_2 = (7 - 8i)$$

Solution

$$\begin{aligned} z_1 \pm z_2 &= (5 + 3i) \pm (7 - 8i) \\ &= (5 \pm 7) + i(3 \pm [-8]) \end{aligned}$$

Multiplication

Multiplication process of two complex numbers is carried out by the normal but taking only in consideration the basic rules defined above, the multiplication process is as follows:

$$\begin{aligned} z_1 \times z_2 &= (x_1 + iy_1) \times (x_2 + iy_2) \\ &= x_1 x_2 + ix_1 y_2 + ix_2 y_1 - y_1 y_2 \end{aligned}$$

Example

Multiply the following two complex numbers:

$$z_1 = (5 + 3i)$$

$$z_2 = (7 - 8i)$$

Solution

$$\begin{aligned} z_1 \times z_2 &= (5 + 3i) \times (7 - 8i) \\ &= 35 - 40i + 21i + 24 \\ &= 59 - 19i \end{aligned}$$

Division

Division of two complex numbers is carried out by multiplying both numerator and denominator by the conjugate of the denominator, as follows:

$$\begin{aligned} \frac{z_1}{z_2} &= \frac{(x_1 + iy_1)}{(x_2 + iy_2)} \\ &= \frac{(x_1 + iy_1)}{(x_2 + iy_2)} \times \frac{(x_2 - iy_2)}{(x_2 - iy_2)} \\ &= \left(\frac{x_1 x_2 - ix_1 y_2 + ix_2 y_1 + y_1 y_2}{x_2^2 + y_2^2} \right) \end{aligned} \tag{20}$$

Example

Divide the following two complex numbers:

$$z_1 = (5 + 3i)$$

$$z_2 = (7 - 8i)$$

Solution

$$\begin{aligned} \frac{z_1}{z_2} &= \frac{(5 + 3i)}{(7 - 8i)} \\ &= \frac{(5 + 3i)}{(7 - 8i)} \times \frac{(7 + 8i)}{(7 + 8i)} \\ &= \frac{11 + 40i}{113} \end{aligned}$$

1.4.1. Commutative Property

Suppose that we have two complex numbers z_1, z_2, then:

$$z_1 + Z_2 = z_2 + Z_1$$
$$z_1 \times Z_2 = z_2 \times Z_1 \tag{21}$$

Example

Verify the associated property for the following two complex numbers:

$$z_1 = (5+3i)$$

$$z_2 = (7-8i)$$

Solution

$$z_1 + z_2 = z_2 + z_1$$

$$\text{R.H.S.} = z_2 + z_1 = (5+3i)+(7-8i) = 12-5i$$

$$\text{L.H.S.} = z_1 + z_2 = (7-8i)+(5+3i) = 12-5i$$

$$\text{R.H.S.} = z_1 \times z_2 = (5+3i)\times(7-8i) = 35-40i+21i+24 = 59-19i$$

$$\text{L.H.S.} = z_2 \times z_1 = (7-8i)\times(5+3i) = 59-19i$$

1.4.2. Associative Property

Suppose that we have three complex numbers z_1, z_2, z_3, then:

$$z_1 + (z_2+z_3) = (z_1+z_2) + Z_3$$
$$z_1 \times (z_2 \times z_3) = (z_1 \times z_2) \times Z_3 \tag{22}$$

1.4.3. Distributive property

Suppose that we have three complex numbers z_1, z_2, z_3, then:

$$z_1 \times (z_2 + z_3) = (z_1 + z_2) \times (z_1 \times z_3) \tag{23}$$

1.5. SPECIAL OPERATION FOR TWO COMPLEX NUMBERS AND THEIR CONJUGATE

1.5.1. Addition

$$\overline{z_1 + z_2} = \overline{z_1} + \overline{z_2} \tag{24}$$

Proof

$$\begin{aligned} L.H.S. = \overline{z_1 + z_2} &= \overline{(x_1 + iy_1) + (x_2 + iy_2)} \\ &= \overline{(x_1 + iy_1)} + \overline{(x_2 + iy_2)} \\ &= (x_1 - iy_1) + (x_2 - iy_2) \\ &= (x_1 + x_2) - i(y_1 + y_2) \end{aligned}$$

$$\begin{aligned} R.H.S. = \overline{z_1} + \overline{z_2} &= \overline{(x_1 + iy_1)} + \overline{(x_2 + iy_2)} \\ &= (x_1 - iy_1) + (x_2 - iy_2) \\ &= (x_1 + x_2) - i(y_1 + y_2) \end{aligned}$$

1.5.2. Multiplication

$$\overline{z_1 \times z_2} = \overline{z_1} \times \overline{z_2} \tag{25}$$

Proof

$$\begin{aligned} L.H.S. = \overline{z_1 \times z_2} &= \overline{(x_1 + iy_1) \times (x_2 + iy_2)} \\ &= \overline{x_1x_2 + ix_1y_2 + ix_2y_1 - y_1y_2} \\ &= \overline{(x_1x_2 - y_1y_2) + i(x_1y_2 + x_2y_1)} \\ &= (x_1x_2 - y_1y_2) - i(x_1y_2 + x_2y_1) \end{aligned}$$

$$LRH.S. = \overline{z_1} \times \overline{z_2} = \overline{(x_1 + iy_1)} \times \overline{(x_2 + iy_2)}$$
$$= (x_1 - iy_1) \times (x_2 - iy_2)$$
$$= x_1 x_2 - i x_1 y_2 - i x_2 y_1 - y_1 y_2$$
$$= (x_1 x_2 - y_1 y_2) - i(x_1 y_2 + x_2 y_1)$$

1.6. DIVISION

$$\overline{\left(\frac{z_1}{z_2}\right)} = \frac{\overline{z_1}}{\overline{z_2}} \; if \; z_2 \neq 0 \tag{26}$$

1.7. REAL PART OF A COMPLEX NUMBER

The real part of the complex number is denoted by $\mathrm{Re}(z)$ and is defined as:

$$\mathrm{Re}(z) = \frac{z + \overline{z}}{2} = \frac{(x + iy) + (x - iy)}{2} = x \tag{27}$$

1.8. IMAGINARY PART OF A COMPLEX NUMBER

The imaginary part of the complex number is denoted by $\mathrm{Im}(z)$ and is defined as:

$$\mathrm{Re}(z) = \frac{z - \overline{z}}{2} = \frac{(x + iy) - (x - iy)}{2} = iy \tag{28}$$

1.9. MODULUS OF THE COMPLEX NUMBER

The absolute value – modulus – of the complex number $z = (x + iy)$ is non negative and real and denoted by $|\,z\,|$ and defined as:

$$|\,z\,| = \sqrt{x^2 + y^2} \tag{29}$$

The absolute value – modulus – of the complex number is the distance between the origin and the point $z = (x, y)$.

Geometric Representation of Complex Number

If

$$z = R(z) + iI(z) = x + iy,$$

then

$$|z| = \sqrt{x^2 + y^2} \tag{30}$$

And its argument or phase angle Θ, see Fig. (**1**) is defined as:

$$\arg(\Theta) = \tan^{-1}\left(\frac{I(z)}{R(z)}\right) \tag{31}$$

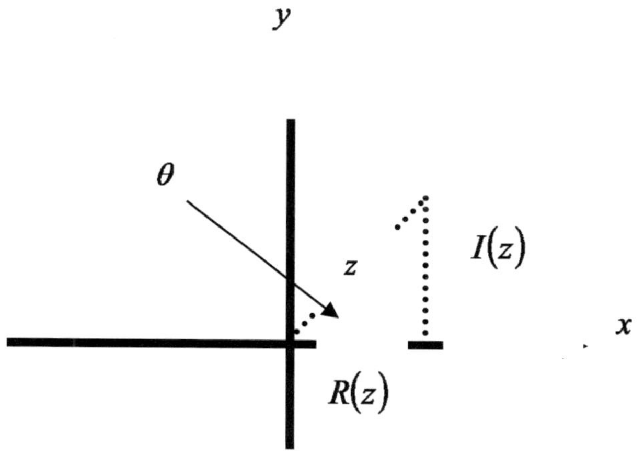

Figure 1: Geometric representation of complex number.

Some Important Rules

$$z + \bar{z} = 2R(z) \tag{32}$$

$$|z| \geq R(z) \tag{33}$$

$$z \times \bar{z} = |z|^2 \tag{34}$$

Example

Find the complex number z , given that $|\,z\,| = 2$ and $\arg(z) = \dfrac{\pi}{4}$

Solution

First let us write the complex number in a general form:

$$z = x + iy$$

We will use the basic ideas for the complex number to solve the given example, let us start with the basic definition of the modulus of the complex number:

$$|\,z\,| = \sqrt{x^2 + y^2} \qquad \text{(A)}$$

Squaring both sides of equation (A), taking into consideration the value of the modulus, leads to:

$$x^2 + y^2 = 4 \qquad \text{(B)}$$

Making use of the $\arg(z) = \dfrac{\pi}{4}$, one can write the following:

$$\arg(z) = \frac{\pi}{4} = \tan^{-1}\left(\frac{y}{x}\right)$$

or

$$\left(\frac{y}{x}\right) = \tan^{-1}\frac{\pi}{4} = 1 \Rightarrow y = x \qquad \text{(C)}$$

Making use of equation (C) into equation (B), leads to:

$$x = \sqrt{2} \qquad \text{(D)}$$

From (c), we get:

$$y = \sqrt{2} \qquad \text{(E)}$$

Using equations (D) and (E) into the complex number, yields:

$$z = \sqrt{2} + i\sqrt{2} \tag{F}$$

Example

$$\text{If } z = \frac{(3+4i)(12-5i)}{(1+i)} \text{ determine } |z|, R(z), \text{ and } I(z)$$

Solution

The first step is to simplify the given complex number, as follows:

$$z = \frac{(3+4i)(12-5i)}{(1+i)} = \frac{(56+33i)}{(1+i)}$$

Simplifying this leads to:

$$z = \frac{1}{2}(89-23i)$$

Now we are ready to get all that was required, as follows:

$$|z| = \sqrt{\left(\frac{89}{2}\right)^2 + \left(\frac{23}{2}\right)^2}$$

$$R(z) = \left(\frac{89}{2}\right)$$

$$I(z) = -\left(\frac{23}{2}\right)$$

Example

Solve the following equation for both x and y:

$$z\bar{z} + (z - \bar{z}) = 0 \tag{A}$$

Solution

Let us re-write both z and $\bar{z}$ in their Cartesian forms as follows:

$$z = x + iy \tag{B}$$

$$\bar{z} = x - iy \tag{C}$$

Now let us substitute from equations (B) and (C) into the left-hand side of equation (A), we get:

$$\textbf{L.H.S.} = (x+iy)(x-iy) + ((x+iy)-(x-iy))$$

$$\textbf{L.H.S.} = x^2 + y^2 + 2iy \tag{D}$$

Now equating real and imaginary parts in equation (D) with zero that is the right-hand side of equation (A), leads to:

$$x^2 + y^2 = 0 \tag{E}$$

$$2y = 0 \Rightarrow y = 0 \tag{F}$$

Therefore, the only solution will be:

$$x = y = 0 \tag{G}$$

1.10. HOLDER INEQUALITIES

The following two inequalities are called Holder inequalities [4]:

$$|z_1 + z_2| \le |z_1| + |z_2| \tag{35}$$

This inequality can be generalized as follows:

$$|z_1 + z_2 + \ldots\ldots + z_n| \le |z_1| + |z_2| + \ldots\ldots\ldots\ldots\ldots + |z_n| \tag{36}$$

$$|z_1 - z_2| \ge \big||z_1| - |z_2|\big| \tag{37}$$

Example

What region in the z-plane is defined by the following inequalities?

(a) $|z-1|+|z+1|<3$

(b) $|z+3|+|z-i|<2$

Solution

Solution of Part (a)

Using Holder's inequality for addition of two complex numbers, lea to:

$$|z_1+z_2| \le |z_1|+|z_2| \tag{A-1}$$

$$|z-1|+|z+1|<3 \tag{A-2}$$

Comparing (A-1) and (A-2), one can get:

$$z_1 = z-1 \tag{A-3}$$

$$z_2 = z+1 \tag{A-4}$$

Now the inequality given in part (a) can then be re-written as follows:

$$|z_1+z_2| \le \left|\underbrace{z-1}_{z_1}\right| + \left|\underbrace{z+1}_{z_2}\right| < 3$$

or

$$|z-1+z+1| \le 3$$

$$|2z| \le 3$$

Then

$$|z| \le \frac{3}{2} \tag{A-5}$$

Equation (A-5) represents an equation of a circle with center at the origin and of radius equals $\frac{3}{2}$

Solution of Part (b)

Using Holder's inequality for addition of two complex numbers, leads to:

$$|z_1 + z_2| \leq |z_1| + |z_2| \qquad \text{(B-1)}$$

$$|z + 3| + |z - i| < 2 \qquad \text{(B-2)}$$

Comparing (B-1) and (B-2), one can get:

$$z_1 = z + 3 \qquad \text{(B-3)}$$

$$z_2 = z - i \qquad \text{(B-4)}$$

Now, we will leave the rest of the solution as a practice for the reader.

Example

Prove that:

$$|z_1 - z_2| \geq \big||z_1| - |z_2|\big|$$

Solution

Start with the solution that deals the left-hand side:

$$\begin{aligned}
\text{L.H.S.} &= |z_1 - z_2|^2 \\
&= (z_1 - z_2)\overline{(z_1 - z_2)} \\
&= z_1\overline{z_1} - z_1\overline{z_2} - \overline{z_1}z_2 + z_2\overline{z_2} \\
&= |z_1|^2 - z_1\overline{z_2} - \overline{z_1}z_2 + |z_1|^2 \\
&= |z_1|^2 + |z_1|^2 - \left(z_1\overline{z_2} + \overline{z_1}z_2\right)
\end{aligned}$$

If we consider

$$\overline{z_1 z_2} = z_1 \overline{z_2}$$

and we know

$$z + \overline{z} = 2R(z)$$

Then, the left-hand side takes the following form:

$$\textbf{L.H.S.} = |z_1|^2 + |z_1|^2 - 2R\left(z_1 \overline{z_2}\right)$$

But

$$|z| \geq R(z)$$
$$\Rightarrow$$
$$2|z| \geq 2R(z)$$

Finally, the left-hand side takes the following form:

$$\textbf{L.H.S.} = |z_1|^2 + |z_1|^2 - 2\left|\overline{z_1} z_2\right|$$

Now then:

$$|z_1 - z_2|^2 \geq |z_1|^2 + |z_2|^2 - 2\left|\overline{z_1} z_2\right|$$
$$\geq \left||z_1| - |z_2|\right|^2$$

or

$$|z_1 - z_2| \geq \left||z_1| - |z_2|\right|$$

1.11. POLAR REPRESENTATION OF THE COMPLEX NUMBER

The complex number $z = (x + iy)$ in polar coordinates takes the following form [5]:

$$z = r(\cos\theta + i\sin\theta) = re^{i\theta} = rcis\theta \qquad (31)$$

In which r represents the distance between the origin and the point $z=(x,y)$, and the angle θ is the angle between the line connecting the origin and the point $z=(x,y)$ and the horizontal axis is measured in anti-clock wise direction.

Example

Express the complex number $z=-5+\sqrt{75}i$ in its polar form.

Solution

To get the required solution, we need to determine both r & θ and to do so, let us follow up the next solution procedure:

$$|z|=\sqrt{x^2+y^2} \tag{A}$$

$$\arg(z)=\tan^{-1}\left(\frac{y}{x}\right) \tag{B}$$

Now, let us use the numerical values given as:

$$|z|=\sqrt{x^2+y^2}=\sqrt{(-5)^2+(\sqrt{75})^2}=10 \tag{C}$$

$$\theta=\arg(z)=\tan^{-1}\left(\frac{\sqrt{75}}{-5}\right)=\tan^{-1}(-\sqrt{3})$$

$$\Rightarrow \tag{D}$$

$$\theta=60^0 \quad \text{or} \quad \theta=120^0$$

See the polar representation in Fig. (**2**).

1.12. ARGUMENT OF THE COMPLEX NUMBER Z

General Case

The argument of the complex number z is denoted by $\arg(z)$ and mathematically defined as:

$$\arg z=\{\theta+2n\pi : z=r(\cos\theta+i\sin\theta),\ n:\text{an integer}\} \tag{38}$$

Note that we write $[\theta \in \arg z]$ as opposed to $[\theta = \arg z]$. We do that because $\arg(z)$ is a set and the designation $\theta \in \arg z$ indicates that θ belongs to that set.

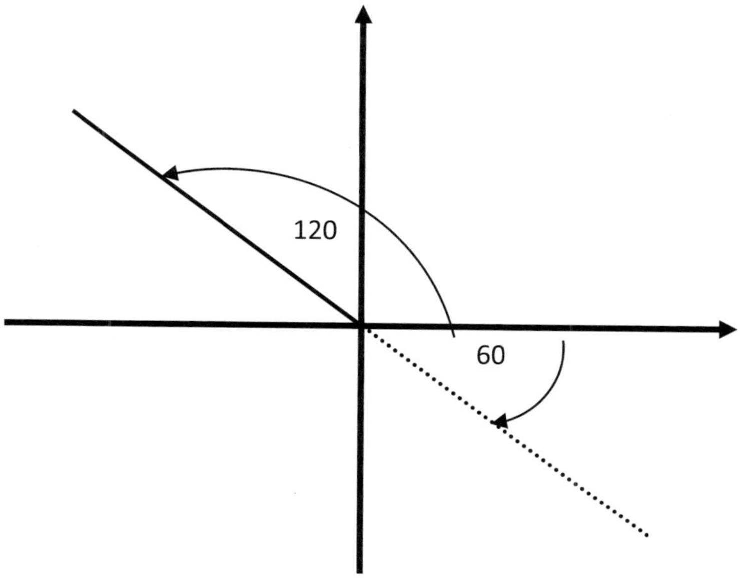

Figure 2: Representation of complex number in polar form.

1.13. MULTIPLICATION OF TWO COMPLEX NUMBERS IN A POLAR FORM

Suppose that we have two complex numbers in their polar forms, and then their product takes the following form:

$$z_1 = r_1(\cos\theta_1 + i\sin\theta_1) = r_1 e^{i\theta_1} = r_1 cis\theta_1 \tag{39}$$

$$z_2 = r_2(\cos\theta_2 + i\sin\theta_2) = r_2 e^{i\theta_2} = r_2 cis\theta_2 \tag{40}$$

Then

$$\begin{aligned} z_1 z_2 &= r_1 r_2 (\cos\theta_1 + i\sin\theta_1)(\cos\theta_2 + i\sin\theta_2) \\ &= r_1 r_2 \left[\cos(\theta_1 + \theta_2) + i\sin(\theta_1 + \theta_2)\right] \\ &= r_1 r_2 e^{i\theta_1} e^{i\theta_2} \\ &= r_1 r_2 cis\theta_1 cis\theta_1 \end{aligned} \tag{41}$$

1.14. DIVISION OF TWO COMPLEX NUMBERS IN A POLAR FORM

Suppose that we have two complex numbers given by equations (39) and (40) in their polar forms, and then their product takes the following form:

$$\frac{z_1}{z_1} = \frac{r_1}{r_2} = \left[\cos\left(\theta_1 - \theta_2\right) + i\sin\left(\theta_1 - \theta_2\right)\right], r_2 \neq 0 \tag{42}$$

Example

The complex number

$$z = 1 + i = \sqrt{2}\left(\cos\frac{\pi}{4} + i\sin\frac{\pi}{4}\right), \text{ then } \arg\left(1+i\right) = \left\{\frac{\pi}{4} + 2n\pi\right\}$$

1.15. ROOTS OF COMPLEX NUMBER

If we return back to:

$$Z^{n} = r^{n}\left(\cos n\theta + i\sin n\theta\right) \tag{43}$$

And if we replace n by $\frac{1}{q}$, then the q th root will be defined as:

$$Z^{\frac{1}{q}} = r^{\frac{1}{q}}\left(\cos\frac{1}{q}\theta + i\sin\frac{1}{q}\theta\right) \tag{44}$$

Equation (38) in a more general form is as follows:

$$Z^{\frac{1}{q}} = r^{\frac{1}{q}}\left(\cos\left(\frac{\theta + 2k\pi}{q}\right) + i\sin\left(\frac{\theta + 2k\pi}{q}\right)\right) \tag{45}$$

In equation (45):

k : Takes an integer value and used to determine the location of the root.

q : Determine the number of the roots

Example

Find five roots of $z = -8i$

Solution

Firstly and to determine the roots, the phase angle θ should be found first as follows:

$$\theta = \tan^{-1}\left[\frac{I(z)}{R(z)}\right] = \tan^{-1}\left[\frac{-8}{0}\right] = \tan^{-1}[-\infty] \Rightarrow \theta = \frac{\pi}{2}$$

$$\tan\theta = \frac{\sin\theta}{\cos\theta}$$

It is important to keep the negative sign in front of the phase angle θ, therefore, if we substitute by $\frac{\pi}{2}$, the sign will be positive but if we use an angle $\frac{3\pi}{2}$, we will get the negative sign. Now we will re-write the formula from which we can find the roots:

$$Z^{\frac{1}{q}} = r^{\frac{1}{q}}\left(\cos\left(\frac{\theta + 2k\pi}{q}\right) + i\sin\left(\frac{\theta + 2k\pi}{q}\right)\right)$$

$$Z^{\frac{1}{5}} = r^{\frac{1}{5}}\left(\cos\left(\frac{\frac{3\pi}{2} + 2k\pi}{5}\right) + i\sin\left(\frac{\frac{3\pi}{2} + 2k\pi}{5}\right)\right)$$

$$Z^{\frac{1}{5}} = r^{\frac{1}{5}}\left(\cos\frac{1}{5}\left(\frac{3\pi}{2} + 2k\pi\right) + i\sin\frac{1}{5}\left(\frac{3\pi}{2} + 2k\pi\right)\right)$$

The next step is the successive substitution by integer values of k starting from zero and stops at four. We will leave this for the reader.

1.16. DRAWING REGIONS IN COMPLEX PLANE

In the present section, we will solve some examples from which the reader will learn how to draw any region in the complex z-plane. This will be helpful when dealing with conformal mapping.

Example

Draw the region $|z| < 1$

Solution

Let us re-write the given equation in a more familiar form, as follows:

$$\sqrt{x^2 + y^2} < 1 \tag{A}$$

Squaring both sides of equation (A), leads to:

$$x^2 + y^2 < 1 \tag{B}$$

Equation (B) represents an equation of the region inside a circle with center at the origin and of radius one unit, see Fig. (**3**).

Example

Draw the region $1 < |z + 2i| \leq 2$

Solution

If we consider the term inside the absolute, and ignore the constant $2i$, we see that the following inequality:

$$1 < |z| \leq 2 \tag{A}$$

will have two circles with common center at the origin and the interior circle with radius greater than 1 unit and the outer circle is of radius less that or equal to 2 units, and so the region between the two circles is represented by equation (A) without any shift. If we take the constant $2i$ into consideration, this means that there will be a shift for the center ***down*** by this amount of that constant, see Fig. (**4**).

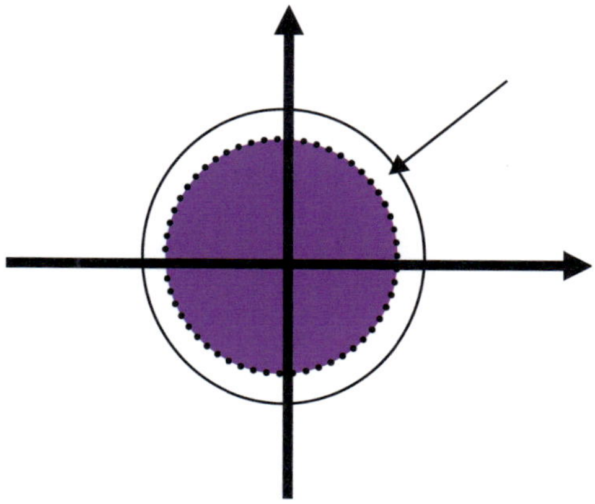

Figure 3: Region $|z| < 1$.

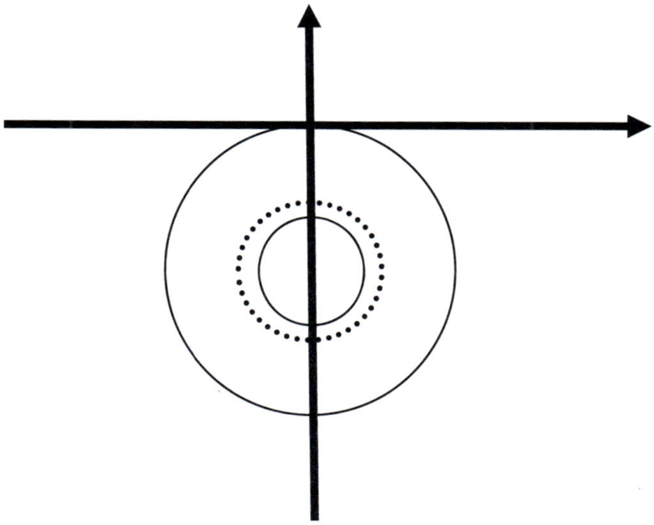

Figure 4: Region $1 < |z + 2i| \leq 2$.

Example

Draw the region $1 \leq |z \pm 2| \leq 3$

Solution

If we consider term inside the absolute, and ignore the constant 2, we see the following inequality:

$$1 \le |z| \le 3 \qquad \text{(A)}$$

will have two circles with common center at the origin and the interior circle with radius greater than or equal to 1 unit and the outer circle is of radius less that or equal to 3 units, and so the region between the two circles is represented by equation (A) without any shift. If we take the constant ± 2 into consideration, this means that there will be a shift for the center ***left for*** $+2$ and ***right for*** -2, see Figs. (**5**) and (**6**), respectively.

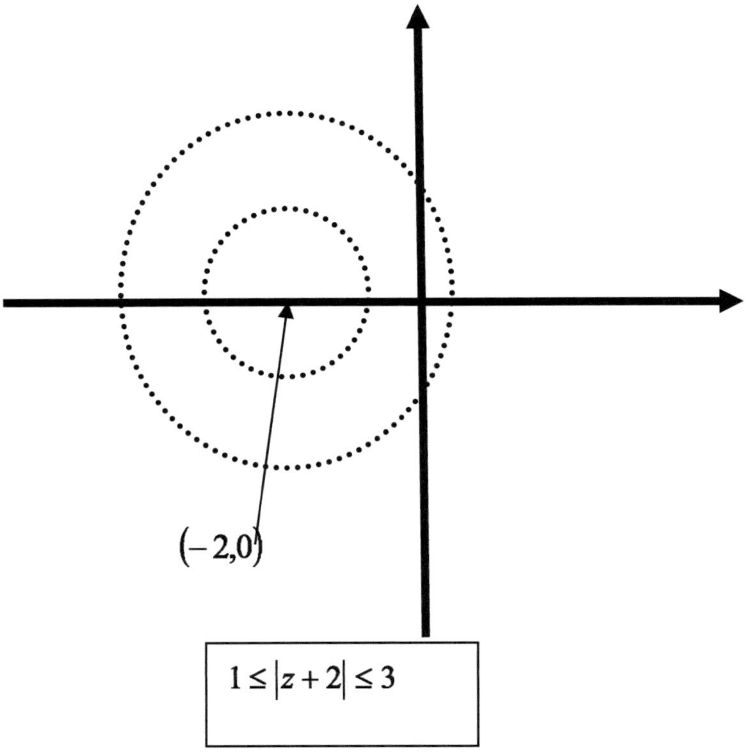

Figure 5: Region $1 \le |z+2| \le 3$.

Exercises

1. Use Bernoulli's technique to get all solutions to the following depressed cubics:

(a) $x^3 - 30x - 36 = 0$

(b) $x^3 - 87x - 130 = 0$

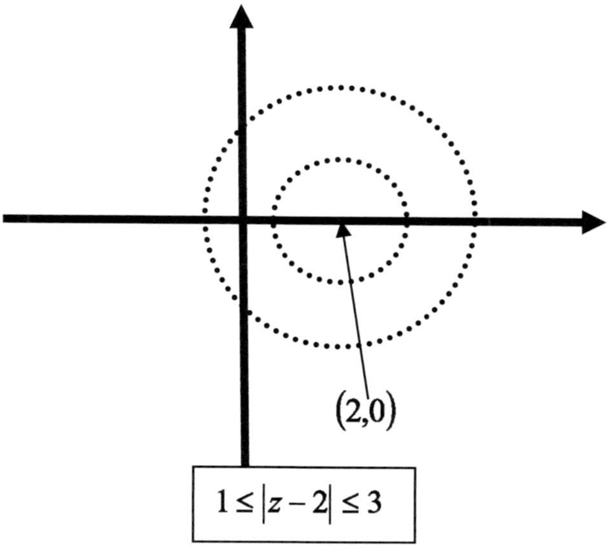

Figure 6: Region $1 \le |z-2| \le 3$.

2. Use Cardano's technique (of substituting $z = x - \frac{a_2}{2}$) to solve the following cubics:

(a) $z^3 - 6z^2 - 3z + 18 = 0$

(b) $z^3 + 3z^2 - 24z + 28 = 0$

3. Perform the required calculations and express your answer in the form $a + ib$:

(a) i^{275}

(b) $i^{\frac{1}{5}}$

(c) $\mathrm{Re}(i)$

(d) $\mathrm{Im}(2)$

(e) $(i-1)^3$

(f) $(7-2i)(3i+5)$

(g) $\mathrm{Re}(7+6i)+\mathrm{Im}(4-4i)$

4. Evaluate the following:

a) $\overline{(1+i)(2+i)}(3+i)$

b) $\dfrac{(3+i)}{\overline{(2+i)}}$

c) $\mathrm{Re}\left[(i-1)^3\right]$

d) $\mathrm{Im}\left[(1+i)^{-2}\right]$

e) Show that $z\bar{z}$ is always a real number.

CHAPTER 2

Complex Functions

Abstract: This chapter is an introduction to the theory of analytic functions of single variable. No one can understand the basic concepts of integration without passing through analytic functions. It will be so difficult starting deeply with such functions but the reader should first start from the basic definitions and then slowly develop the ideas of complex analysis to the point that considered the bases to deeper topics such as Cauchy's theorem and the Riemann mapping theorem.

Keywords: Complex-valued function, domain and range of complex-valued function, Cauchy theorem, Riemann mapping theorem, mapping, linear transformation, inverse transformation, bilinear transformation, logarithmic complex function.

2.1. THE COMPLEX FUNCTION-DEFINITION

The complex function is a relation between the complex number in z-plane and the corresponding image in the w-plane. This relation may have different forms, linear, quadratic, bilinear, etc. and is usually written as $w = f(z)$ **[6]**.

2.2. THE DOMAIN AND THE RANGE

Assume that we have a non-empty set D of all possible values in the complex plane z and under certain prescribed rule that assign to each value of $z \in D$ one and only one another complex number w, then the non-empty set D is the domain of the complex function. Now then one can define the range as the set of all images $\{w = f(z) : z \in D\}$.

Definition: Mapping – Transformation –

Mappings or transformations is a process of transferring point, line, area or something like that from z-plane to another w-plane. One can write the mapping or transformation in its general case as follows:

$$w = f(z) = f(x + iy) = u + iv \qquad (1)$$

Both u and v can then written as:

$$u = u(x, y) \ \ \& \ \ v = v(x, y) \qquad (2)$$

S. G. Ahmed

Taking into consideration equations (1-2), then the complex function can take the following form, Fig. **1**:

$$f(z) = f(x+iy) = u(x,y) + iv(x,y) \tag{3}$$

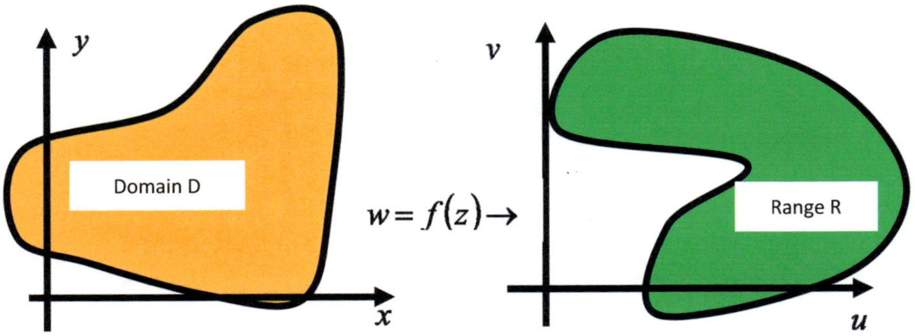

Figure 1: Domain and range.

Example

Write $f(z) = z^4$ in the form $f(z) = u(x,y) + iv(x,y)$

Solution

Expanding the given function using the binomial formula, leads to:

$$f(z) = (x+iy)^4 = \left(x^4 - 6x^2y^2 + y^4\right) + i\left(4x^3y - 4xy^3\right)$$

So that:

$$v(x,y) = \left(4x^3y - 4xy^3\right)$$

Example

Express by a formula involving the variables z and $\bar{z}$

Solution

We know that:

$$x = \left(\frac{z+\bar{z}}{2}\right)$$

and

$$y = \left(\frac{z - \bar{z}}{2i} \right)$$

Substituting into original function, leads to:

$$f(z) = 4\left(\frac{z + \bar{z}}{2} \right)^2 + i4\left(\frac{z - \bar{z}}{2i} \right)^2$$

Simplify, leads to:

$$f(z) = (1 - i)z^2 + (2 + 2i)z\bar{z} + (1 - i)\bar{z}^2$$

Now, let us make use of:

$$w(u, v) = u(x, y) + iv(x, y) = f(z)$$

The function $w(u, v)$ is called the transformation as shown in Fig. (**2**).

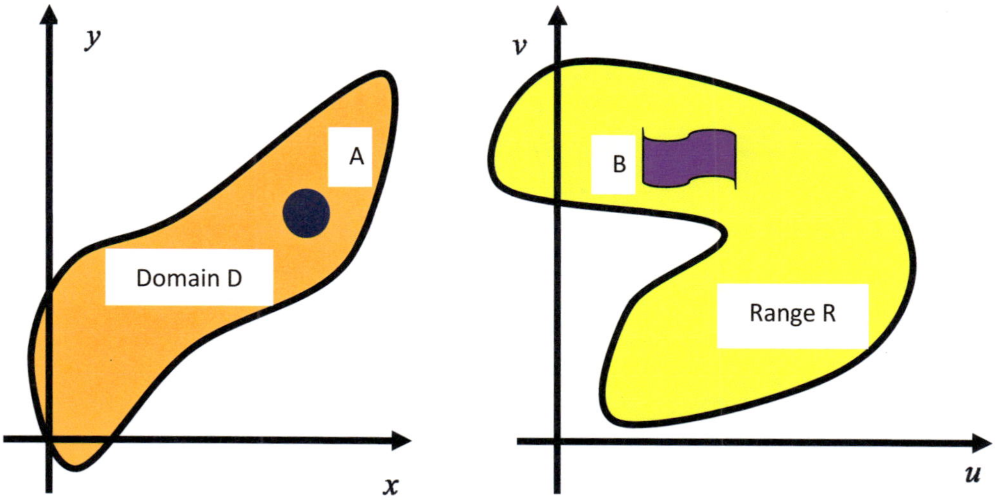

Figure 2: f maps A into B .

One-to-One Function

A function $f(z)$ is said to be one-to-one if it maps different points in the z-plane onto corresponding different points in the w-plane **[7]**.

Example

Under the transformation $w = iz$ verify that the line $y = x + 1$ in the z-plane will be another line $v = -u - 1$ in the w-plane.

Solution

Let A be the set of all points satisfying the equation of the straight line $y = x + 1$ in the z-plane as follows:

$$A = \{(x, y) : y = x + 1\}$$

We looking for a function

$$B = f(A)$$

To do so, let $z = x + iy \in A$

Now, for any point $z_o \in A$, there exists a corresponding point $w_o \in B$

Therefore

$$w_o = f(z_o) \text{ iff } f^{-1}(w_o) = z_o$$

Now then we can write the following mathematical manipulation

$$u + iv = w = f(z) \in B \Leftrightarrow f^{-1}(w) = z = x + iy \in A$$
$$\Leftrightarrow -iw \in A$$
$$\Leftrightarrow v - iu \in A$$

$$u + iv = w = f(z) \in B \Leftrightarrow (v, -u) \in A$$
$$\Leftrightarrow -uv + 1$$
$$\Leftrightarrow v = -u - 1$$

2.3. ELEMENTARY MAPPINGS

2.3.1. Translation

To understand the translation, let we have square shown in Fig. (**3**) and let $B = a + ib$ be a fixed complex constant, then the translation can be defined as:

$$w = T(z) = z + B = x + a + i(y + b) \tag{4}$$

Equation (4) is a one-to-one mapping from z-plane onto w-plane **[8]**. The inverse mapping is given by:

$$z = T^{-1}(w) = w - B = u - a + i(v - b) \tag{5}$$

equation (5) shows that T is one-to-one mapping from z-plane onto the w-plane.

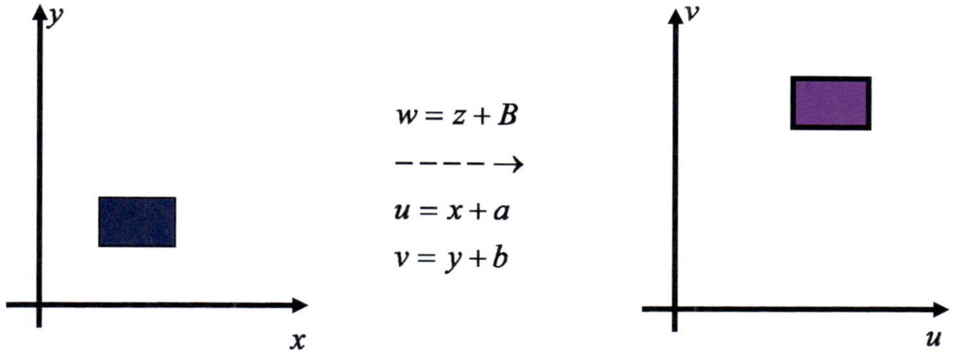

Figure 3: The transformation $w = T(z)$.

2.3.2. Rotation

To understand the concept of rotation, let us have a fixed real number say α, therefore, for any any complex number $z = e^{i\theta}$, the rotation can be defined as:

$$w = R(z) = ze^{i\alpha} = re^{i(\alpha + \alpha)} \tag{6}$$

Equation (6) is a one-to-one mapping and is called a rotation **[9]**. The inverse mapping is:

$$z = R^{-1}(w) = we^{-i\alpha} = \rho e^{i(\phi - \alpha)} \tag{7}$$

The effect of rotation is shown in Fig. (**4**).

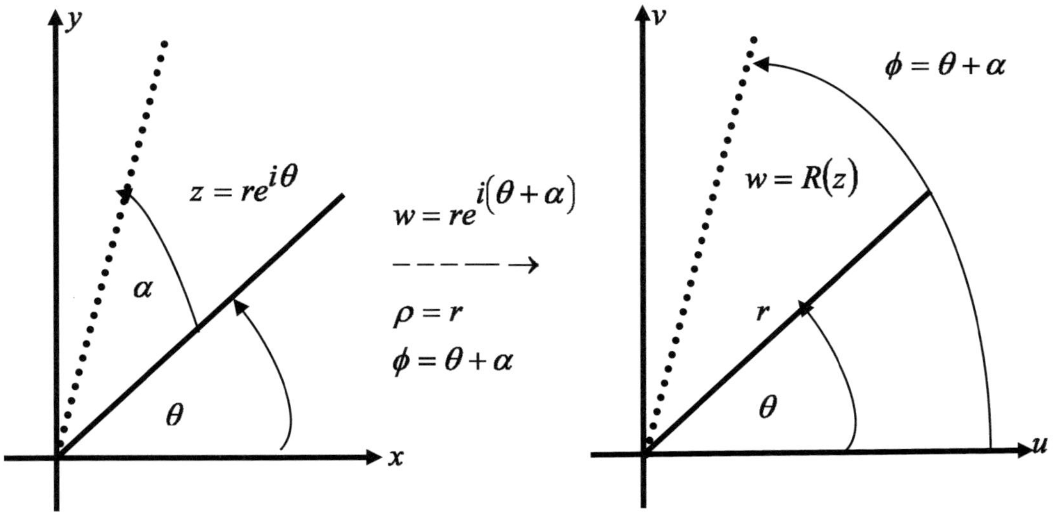

Figure 4: The rotation $w = R(z) = ze^{i\alpha} = re^{i(\alpha+\alpha)}$.

Example

The parametric equation of an ellipse with major axis of length 4 units and minor axis of length 2 units take the following form:

$$s(t) = 2\cos t + i\sin t = (2\cos t, \sin t), \text{ for } 0 \le t \le 2\pi$$

If we rotate the given ellipse by an angle $\frac{\pi}{6}$ and shift the center two units right and one unit up, mathematically both previous operations can be carried out through the following steps:

$$r(t) = s(t)e^{i\frac{\pi}{6}} + (2+i)$$

$$r(t) = (2\cos t + i\sin t)\left(\cos\frac{\pi}{6} + i\sin\frac{\pi}{6}\right) + (2+i)$$

$$r(t)=\left(2\cos t\cos\frac{\pi}{6}-\sin t\sin\frac{\pi}{6}\right)+i\left(2\cos t\sin\frac{\pi}{6}+\sin t\cos\frac{\pi}{6}\right)+(2+i)$$

$$r(t)=\left(\sqrt{3}\cos t-\frac{1}{2}\sin t\right)+i\left(\cos t+\frac{\sqrt{3}}{2}\sin t+1\right)$$

$$r(t)=\left(\sqrt{3}\cos t-\frac{1}{2}\sin t+2\right)+i\left(\cos t+\frac{\sqrt{3}}{2}\sin t+1\right)$$

$$r(t)=\left(\sqrt{3}\cos t-\frac{1}{2}\sin t+2\ ,\ \cos t+\frac{\sqrt{3}}{2}\sin t+1\right),$$

for $0\le t\le 2\pi$

2.3.4. Magnification

For any fixed and real number k, the magnification can be defined through the following equation:

$$w=S(z)=Kz=Kx+iKy \tag{8}$$

Equation (8) is a one-to-one mapping of the z-plane onto the w-plane. The fixed constant may take one of the following two cases:

Case (1): If $k>1$ it will has the effect of increasing the distance between the corresponding points.

Case (2): If $k<1$ it will has the effect of decreasing the distance between the corresponding points.

The inverse transformation is given by:

$$z=S^{-1}(w)=\frac{1}{K}w=\frac{1}{K}u+i\frac{1}{K}v \tag{9}$$

The effect of magnification is shown in Fig. (**5**).

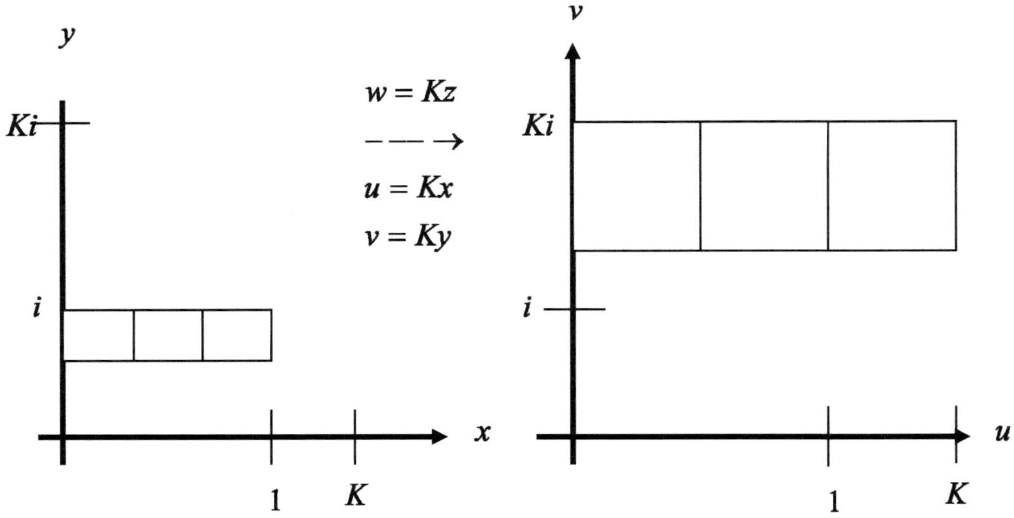

Figure 5: The magnification $w = Kx + iKy$.

2.4. Linear Transformation

To understand the concept of linear transformation, assume that we have:

$$A = Ke^{i\alpha} \tag{10-1}$$

and

$$B = a + ib \tag{10-2}$$

The constant k in equation (10-1) is a positive real number, combining equations (10-1,2) as follow:

$$w = L(z) = Az + B \tag{10-3}$$

Then, equation (10) represents a one-to-one mapping of the z-plane onto the w-plane and is called a linear transformation **[10]**. The inverse mapping is given by:

$$z = L^{-1}(w) = \frac{1}{A}w - \frac{B}{A} \tag{11}$$

Example

Given a linear transformation $w = iz + i$, show that the given transformation maps the right-half plane $\text{Re}(z) \geq 1$ onto the upper-half plane $\text{Im}(w) \geq 2$

Solution

Let A be the set of all points in the right-half plane $\text{Re}(z) \geq 1$ defined as follows:

$$A = \{(x, y) : x \geq 1\}$$

We assumed this set to describe the following:

$$B = f(A)$$

Solve $w = iz + i$ for z, leads to:

$$z = \frac{w - i}{i} = -iw - 1 = f^{-1}(w)$$

Subsequently

$$f^{-1}(w) = -iw - 1$$

Now suppose that:

$$z_o \in A$$

Then

$$w_o \in B$$

Therefor

$$w_o = f(z_o) \ \textit{iff} \ f^{-1}(w_o) = z_o$$

We have:

$$u + iv = w = f(z) \in B \Leftrightarrow f^{-1}(w) = z \in A$$
$$\Leftrightarrow -iw - 1 \in A \Leftrightarrow -v - 1 - iu \in A$$
$$\Leftrightarrow (v - 1, -u) \in A \Leftrightarrow v - 1 \geq 1 \Leftrightarrow v \geq 2$$

Thus $B = \{w = u + iv : v \geq 2\}$

which is the same as saying $\text{Im}(w) \geq 2$. The result of this example is shown in Fig. (**6**).

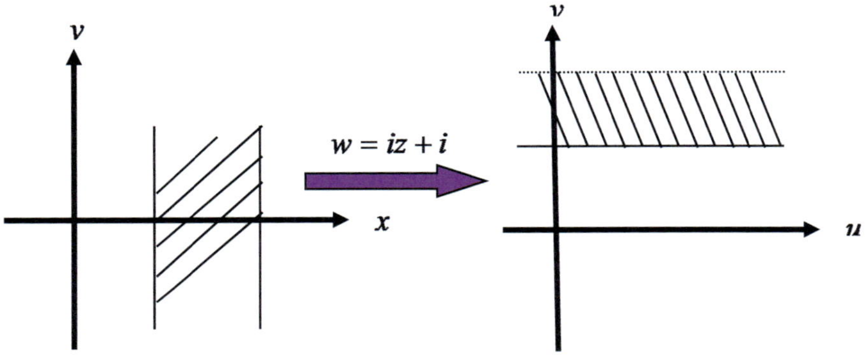

Figure 6: The linear transformation $w = iz + i$.

2.5. Elementary Functions

We will start this section by a simple question, ***How the complex-valued function like*** $e^z, \log z, \sin z$ ***should be defined?*** to have answers to this question, the following points should be covered:

1. When z is a pure real, these functions should give the same values as the corresponding function of real variables, for example, if z is a pure real, then e^Z will give the same value as e^Z
2. The properties of these functions should correspond the function of real variables, for example, $e^{Z_1 + Z_2} = e^{Z_1} e^{Z_2}$ to be valid regardless of whether z were real or complex **[11]**.

2.6. COMPLEX EXPONENTIAL FUNCTION

In the real analysis, the real exponential function and its expansion are as follow:

$$e^x = \exp(x) = \sum_{n=0}^{\infty} \frac{1}{n!} x^n$$

In a correspondence, the complex exponential function is then defined as:

$$e^z = \exp(z) = \sum_{n=0}^{\infty} \frac{1}{n!} z^n$$

It is clear that there is an agreement between the two expansions, when z is a pure real **[12]**.

Theorem

The complex exponential function should satisfy the following conditions:

1. $\frac{d}{dz}\exp(z) = \exp(z)$

2. $\exp(z_1 + z_2) = \exp(z_1)\exp(z_2)$

3. If θ is real number, then $e^{i\theta} = \cos\theta + i\sin\theta$

We often use $\exp(z)$ to denote the single-valued exponential function. Some additional properties of $\exp(z)$:

1. $e^{z+i2n\pi} = e^z$ for all z, provided n is an integer

2. $e^z = 1$ iff $z = i2n\pi$, where n is an integer

3. $e^{z_1} = e^{z_2}$ iff $z_2 = z_1 + i2n\pi$, for some integer n.

Example

For any integer n, the points

$$z_n = \frac{5}{4} + i\left(\frac{11\pi}{6} + 2n\pi\right)$$

in the z-plane are mapped onto the single point:

$$\begin{aligned} w_o &= \exp(z) \\ &= e^{\frac{5}{4}} + \left(\cos \frac{11\pi}{6} + i \sin \frac{11\pi}{6} \right) \\ &= \frac{\sqrt{3}}{2} e^{\frac{5}{4}} - i \frac{1}{2} e^{\frac{5}{4}} \\ &\cong 3.02 - 1.75i \end{aligned}$$

in the w-plane, as indicated in Fig. (7).

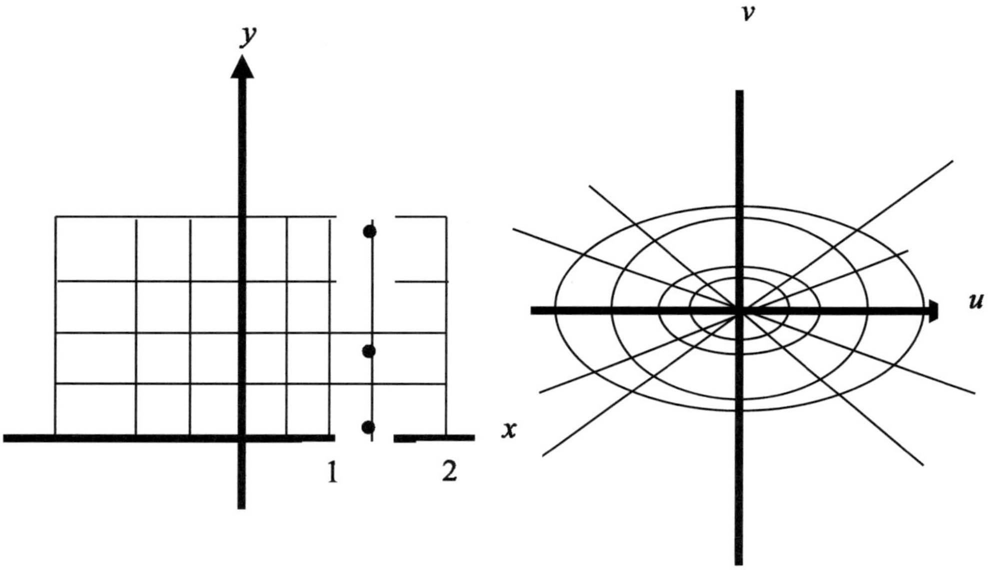

Figure 7: The points z_n in the z-plane and their image w_0 in the w-plane.

2.6.1. Range of Exponential Function

The complex exponential function $e^z = e^x(\cos y + i \sin y)$ cannot equal zero, since the exponential part is never zero and both sine and cosine are never zero.

Suppose that $w = e^z \neq 0$ and if we write w in its exponential form as $w = \rho e^{i\varphi}$ then:

$$\rho e^{i\phi} = e^x e^{iy}$$

Using

$$e^{z} = e^{x}e^{iy} = e^{x}\left(\cos y + i\sin y\right)$$

and

$$z_1 = z_2 \Leftrightarrow r_1 e^{i\theta_1} = r_2 e^{i\theta_2} \quad \text{iff } r_1 = r_2 \ \& \ \theta_1 = \theta_2 + 2\pi k$$

We get:

$\rho = e^{x}$ & $\phi = y + 2n\pi$, where n is an integer.

Therefore:

$$\rho = \left|e^{x}\right| = e^{x}$$

$$\varphi \in \arg\left(e^{z}\right) = \left\{\operatorname{Arg}\left(e^{z}\right) + 2n\pi\right\}$$

Let us solve ρ and ϕ for x and y, leads to:

$$x = \ln \rho$$
$$y = \varphi + 2n\pi$$

From the above two equations, one can write:

$$\begin{aligned} z &= x + iy \\ &= \ln \rho + i(\phi + 2n\pi) \\ &= \ln|w| + i(\operatorname{Arg} w + 2n\pi) \end{aligned}$$

Hence

$$\exp\left(\ln|w| + i(\operatorname{Arg} w + 2n\pi)\right) = w$$

From the above analysis one can conclude that the transformation $w = e^{z}$ maps the complex plane onto the set of nonzero complex numbers.

2.7. THE COMPLEX LOGARITHM

If the logarithmic complex function is a non-zero, then as derived before, it will has two solutions, and since e^Z is many-to-one, therefore its inverse will be multi-valued. This will lead to the following definition:

Definition

For a non-zero complex number, $z \neq 0$, the multi-valued function log is defined as the inverse of the exponential function; that is:

$$\log(z) = w \quad \text{iff} \quad z = \exp(w) \tag{12}$$

If we go through the same steps as we did in:

$$\begin{aligned} x &= \ln \rho \\ y &= \varphi + 2n\pi \end{aligned} \tag{13}$$

and

$$z = \ln|w| + i(\text{Arg}w + 2n\pi) \tag{14}$$

$$\exp(\ln|w| + i(\text{Arg}w + 2n\pi)) = w \tag{15}$$

We find that, for any complex number $z \neq 0$, the solution w takes the form:

$$\log(z) = w \quad \text{iff} \quad z = \exp(w)$$

or

$$w = \ln(z) + i\theta \quad , \; z \neq 0 \tag{16}$$

Before, $\arg(z)$ is defined as:

$$\arg(z) = \{\text{Arg}(z) + 2n\pi : n \,\text{is an integer}\}$$

Therefore, we can write an expression for the set of all values of log (z) as follows:

$$\log(z) = \{\ln|z| + i(\text{Arg}(z) + 2n\pi : n \text{ is an integer})\}$$
$$= \ln|z| + i\arg(z) \tag{17}$$

2.8. THE PRINCIPLE VALUE

Herein, we will turn our attention to the principle value of the logarithm and to do so let for a non-zero complex number, the principle value of the logarithm will be defined as **[13]**:

$$\text{Log}(z) = \ln|z| + i\text{Arg}(z) \tag{18}$$

Example

Find the values of $\text{Log}(1+i)$ & $\text{Log}(i)$

Solution

Let us start the solution by re-writing the following relation:

$$\text{Log}(z) = \ln|z| + i\text{Arg}(z)$$

$$\text{Log}(1+i) = \ln|(1+i)| + i\text{Arg}(1+i) = \ln\sqrt{2} + i\left(\frac{\pi}{4} + 2n\pi\right)$$

$$\text{Log}(i) = \ln|i| + i\text{Arg}(i) = i\left(\frac{\pi}{2} + 2n\pi\right)$$

The principle values are:

$$\text{Log}(1+i) = \ln\sqrt{2} + i\left(\frac{\pi}{4} + 2n\pi\right) = \frac{\ln 2}{2} + i\frac{\pi}{4}$$

$$\text{Log}(i) = i\left(\frac{\pi}{2} + 2n\pi\right) = i\frac{\pi}{2}$$

We now investigate some of the properties of log & Log. From the previous equations, it follows that:

$$\exp(\text{Log } z) = z \quad \textit{for all } z \neq 0$$

$$\text{Log}(\exp z) = z \quad \text{Pr}\textit{ovided } -\pi < \text{Im}(z) \leq \pi$$

The following example illustrates that, even though Log is not continuous along the negative real axis, it is still defined there. Let us remained the reader to the fact that even though the log is not continuous along the negative real axis, it still defined over there, the next example will clarify this fact.

Example

Check if the identity $\text{Log}(z_1 z_2) = \text{Log}(z_1) + \text{Log}(z_2)$ is not always valid.

Solution

Assume that we have the following two complex numbers:

$$z_1 = -\sqrt{3} + i$$
$$z_2 = -1 + i\sqrt{3}$$

Then

$$\begin{aligned} \text{Log}(z_1 z_2) &= \log[(-\sqrt{3} + i)(-1 + i\sqrt{3})] \\ &= \text{Log}(-4i) \end{aligned}$$

The result in the right-hand side can be re-written as:

$$\text{Log}(z_1 z_2) = \ln 4 + i\left(-\frac{\pi}{2}\right)$$

Now let deal with the sum of two logarithmic numbers:

$$\text{Log}(z_1) + \text{Log}(z) = \ln 2 + i\frac{5\pi}{6} + \ln 2 + i\frac{2\pi}{3} = \ln 4 + i\frac{3\pi}{2}$$

Theorem

The identity $\text{Log}(z_1 z) = \text{Log}(z_1) + \text{Log}(z)$ holds true iff

$$-\pi < \mathrm{Arg}\left(z_1\right)+\mathrm{Arg}\left(z\ \right) < \pi \tag{19}$$

Theorem

Let z_1 & z_2 be non-zero complex numbers. The multi-valued function log satisfy the same properties of the logarithms:

Addition

$$\log\left(z_1 z_2\right)=\log\left(z_1\right)+\log\left(z_2\right) \tag{20}$$

Subtraction

$$\log\left(\frac{z_1}{z_2}\right)=\log\left(z_1\right)-\log\left(z_2\right) \tag{21}$$

Reciprocal

$$\log\left(\frac{1}{z}\right)=-\log\left(z\right) \tag{22}$$

2.9. GENERAL SOLVED EXAMPLES

Example

Show that $R(z)\geq 0$ implies $I(w)\leq 0$ for $w=\left(\dfrac{3z+i}{i-z}\right)$

Solution

Let start the solution by dealing with the given function w as follows:

$$\begin{aligned} w &= \left(\frac{3z+i}{i-z}\right) \\ &= \left(\frac{3(x+iy)+i}{i-(x+iy)}\right) \\ &= \left(\frac{3x+i(3y+1)}{-x+i(1-y)}\right)\times\left(\frac{-x-i(1-y)}{-x-i(1-y)}\right) \end{aligned}$$

Simplification leads to:

$$w=\left(\frac{-3x^2-3y^2+2y+1}{x^2+(1-y)^2}\right)+i\left(\frac{-4x}{x^2+(1-y)^2}\right)$$

Note the following:

The term $\left(\frac{-3x^2-3y^2+2y+1}{x^2+(1-y)^2}\right)$ represents $R(w)$

The term $\left(\frac{-4x}{x^2+(1-y)^2}\right)$ represents $I(w)$

The numerator of $\left(\frac{-4x}{x^2+(1-y)^2}\right)$ without the negative sign represents $R(z)$

Therefore;

If $R(z)\geq 0 \Rightarrow I(w)\leq 0$

Example

Show that $I(z)\geq 0$ implies $I(w)\leq 0$ for $w=\left(\frac{z+3}{z+2}\right)$

Solution

Let start the solution by dealing with the given function w as follows:

$$\begin{aligned}w&=\left(\frac{z+3}{z+2}\right)\\&=\left(\frac{(x+iy)+3}{(x+iy)+2}\right)=\left(\frac{(x+3)+iy}{(x+2)+iy}\right)\\&=\left(\frac{(x+3)+iy}{(x+2)+iy}\right)\times\left(\frac{(x+3)-iy}{(x+2)-iy}\right)\\&=\underbrace{\left(\frac{x^2+5x+6+y^2}{(x+2)^2+y^2}\right)}_{R(w)}+\underbrace{\left(\frac{-i\overbrace{(y)}^{I(z)}}{(x+2)^2+y^2}\right)}_{I(w)}\end{aligned}$$

Therefore;

$$\text{If}\, I(z) \geq 0 \Rightarrow I(w) \leq 0$$

Example

$$\text{If}\, w = \frac{i(1-z)}{(1+z)},\ \text{show that if}\, |z| < 1\, \text{implies}\, I(w) > 0$$

Solution

Let start the solution by dealing with the given function w as follows:

$$\begin{aligned}
w &= \left(\frac{i(1-z)}{(1+z)}\right) \\
&= \frac{i(1-(x+iy))}{(1+(x+iy))} \\
&= \frac{y+i(1-x)}{(1+x)+iy} \times \frac{(1+x)-iy}{(1+x)-iy} \\
&= \underbrace{\frac{2y}{(x^2+y^2)+1}}_{R(w)} + i\underbrace{\left\{-\frac{\left[\overbrace{(x^2+y^2)}^{|z|}\right]+1}{(x^2+y^2)+1}\right\}}_{I(w)}
\end{aligned}$$

As we see, the imaginary part of w is multiplied by negative sign, therefore;

$$\text{if}\, |z| < 1\, \text{implies}\, I(w) > 0.$$

Example

If $w = z + \dfrac{1}{z}$, what restriction must be imposed on z to ensure $I(w) > 0$

Solution

Let start the solution by dealing with the given function w as follows:

$$w=\left(z+\frac{1}{z}\right)=\left(\frac{z^2+1}{z}\right)$$

$$=\left(\frac{(x+iy)^2+1}{(x+iy)}\right)$$

$$=\left(\frac{(x+iy)^2+1}{(x+iy)}\right)\times\left(\frac{(x-iy)}{(x-iy)}\right)$$

$$=\underbrace{\frac{x(x^2-y^2)+2xy^2+x}{x^2+y^2}}_{R(w)}+i\underbrace{\left\{\frac{(2x^2y-x^2y+y^3-y)}{x^2+y^2}\right\}}_{I(w)}$$

Now to get a restriction to ensure that $I(w)>0$ let us deal with $R(w)>0$ as follows:

$$R(w)=\frac{x(x^2-y^2)+2xy^2+x}{x^2+y^2}>0$$

$$=\frac{y(x^2+y^2-1)}{x^2+y^2}>0\Rightarrow$$

$$y\left(\underbrace{x^2+y^2}_{|z|^2}-1\right)>0$$

Therefore, to ensure that $I(w)>0$ will be positive for any case of both x, y the term $|z|^2>1$ and that will be the required condition.

2.10. CAUCHY-RIEMANN EQUATIONS

A complex function $w=u+iv$ is said to be analytic if it satisfies the following Cauchy-Riemann equations **[14]**:

$$\frac{\partial u}{\partial x} = \frac{\partial v}{\partial y} \tag{23}$$

$$\frac{\partial u}{\partial y} = -\frac{\partial v}{\partial x} \tag{24}$$

2.10.1. Definition: Harmonic Complex Function

A complex function $w = u + iv$ is said to be harmonic if the second derivatives of u and v with respect to the spatial variables x and y exist [15].

Example

Construct an analytic function $v = 2x(1-y)$

Solution

Let us start the solution by re-writing the Cauchy-Riemann equations as follows:

$$\frac{\partial u}{\partial x} = \frac{\partial v}{\partial y} \tag{A}$$

$$\frac{\partial u}{\partial y} = -\frac{\partial v}{\partial x} \tag{B}$$

Next, is to find the partial derivative of v with respect to the spatial variable y leads to:

$$\frac{\partial v}{\partial y} = -2x \tag{C}$$

Then

$$\frac{\partial u}{\partial x} = -2x \tag{D}$$

Integrating both sides of equation (D), leads to:

$$u = -x^2 + \Phi(y) \tag{E}$$

Again but this time we will deal with the second Cauchy-Riemann equation given by equation (B), as follows:

$$\frac{\partial v}{\partial x} = (-)[2(1-y)] \tag{F}$$

Integrating both sides of equation (F) with respect to the spatial variable x, we get:

$$u = -2x + 2xy + \Psi(x) \tag{G}$$

Now then, the final form of the function u comes from adding equation (E) and (G) taking into consideration that the repeated term is taken once time only, therefore;

$$u = -x^2 - 2x + 2xy \tag{H}$$

Note that, we already got the function u and the function v already given, so the function $w = u + iv$ will take the following form:

$$w = u + iv = \left(-x^2 - 2x + 2xy\right) + i\left(2x(1-y)\right) \tag{I}$$

We can say from this example, that if v is given, the procedure to find w is to differentiate v, then use Cauchy-Riemann equation and integrate the partial derivative of u times the number of the spatial variables, and vice versa when u is given.

Example

Find $w(z)$ given $w'(z) = 4z - 3$ and $w(1+i) = -3i$

Solution

Start the solution by integrating the derivative given in the problem, as follows:

$$\int \frac{dw}{dz} = \int (4z-3)dz$$

$$w(z) = 2z^2 - 3z + C \quad \text{(A)}$$

Next is make use of the given condition to get the value of the constant of integration, leads to:

$$w(1+i) = 2(1+i)^2 - 3(1+i) + C = -3i \quad \text{(B)}$$

From equation (B), one can get the value of the constant of integration and subsequently, the required function.

Example

Find $v(x, y)$ if $u(x, y) = e^x \cos y$ then show that $w(u, v)$ is a harmonic function.

Solution

We previously knew that a function is said to be harmonic if the second derivatives of u and v with respect to the spatial variables x and y exist as well as it satisfies the Cauchy-Riemann equations. As we see only the function $u(x, y)$ is given, so we need to find first the second function $v(x, y)$, this will be the first step. Then the next step is to check the validity of Cauchy-Riemann equations, finally the third step is to check the existence of the second derivatives. To find the function $v(x, y)$, let follow the following procedure;

$$\frac{\partial u(x, y)}{\partial x} = e^x \cos y \quad \text{(A)}$$

$$\frac{\partial u(x, y)}{\partial y} = -e^x \sin y \quad \text{(B)}$$

Re-write the Cauchy-Riemann equations:

$$\frac{\partial u}{\partial x} = \frac{\partial v}{\partial y} \quad \text{(C)}$$

$$\frac{\partial u}{\partial y} = -\frac{\partial v}{\partial x} \quad \text{(D)}$$

Substituting from (A) into (C), leads to:

$$\frac{\partial v}{\partial y} = e^x \cos y \tag{E}$$

Integrating both sides of equation (E), gives:

$$v(x, y) = e^x \sin y + \Phi(x) \tag{F}$$

Substituting from (B) into (D), leads to:

$$\frac{\partial v}{\partial x} = e^x \sin y \tag{G}$$

Integrating both sides of equation (G), gives:

$$v(x, y) = e^x \sin y + \Psi(y) \tag{H}$$

Adding equations (F) and (H), leads to:

$$v(x, y) = e^x \sin y \tag{I}$$

Now the function $w(u, v)$ to be harmonic if $u(x, y)$ and $v(x, y)$ satisfy the following equations:

$$\frac{\partial^2 u(x, y)}{\partial x^2} + \frac{\partial^2 u(x, y)}{\partial y^2} = 0$$

Or

$$\frac{\partial^2 v(x, y)}{\partial x^2} + \frac{\partial^2 v(x, y)}{\partial y^2} = 0$$

We will leave finding the derivatives to the readers.

Example

Find all solutions for $\sin z = 3$.

Solution

First of all, let us express the sine function in terms of complex exponential function therefore; the given equation takes the following form:

$$\frac{e^{iz} - e^{-iz}}{2i} = 3 \qquad \text{(A)}$$

Multiplying both sides of equation (A) by e^{iz} to remove the corresponding one with negative power, yields:

$$e^{iz}\, e^{iz} - 1 = 6ie^{iz} \qquad \text{(B)}$$

Simplifying equation (B), leads to:

$$e^{2iz} - 6ie^{iz} - 1 = 0 \qquad \text{(C)}$$

Equation (C) can take the following form:

$$\left(e^{iz}\right)^2 - 6i\left(e^{iz}\right) - 1 = 0 \qquad \text{(D)}$$

Equation (D) is a quadratic equation; its solution takes the following form:

$$\left(e^{iz}\right) = 3 \pm \frac{1}{2}\sqrt{40} \qquad \text{(E)}$$

Taking the natural logarithm for both sides of equation (E), yields:

$$z = i\ln\left(3 \pm \frac{1}{2}\sqrt{40}\right) \qquad \text{(F)}$$

Note that $\sqrt{40} > 3$ and due to the basic definition of the logarithmic function that it is defined on the positive side only so; equation (F) is then takes the following form:

$$z = i\ln\left(3 + \frac{1}{2}\sqrt{40}\right) \qquad \text{(G)}$$

Example

Find all solutions for $\cosh z = -2$.

Solution

First of all, let us express the $\cosh z$ function in terms of exponential function therefore; the given equation takes the following form:

$$\frac{e^{z} + e^{-z}}{2} = -2 \tag{A}$$

Multiplying both sides of equation (A) by e^{z} to remove the corresponding one with negative power, yields:

$$e^{z} e^{z} + 1 = -4e^{iz} \tag{B}$$

Simplifying equation (B), leads to:

$$e^{2z} + 4e^{z} = 1 \tag{C}$$

Equation (C) can take the following form:

$$\left(e^{z}\right)^{2} + 4\left(e^{z}\right) = 1 \tag{D}$$

Equation (D) is a quadratic equation; its solution takes the following form:

$$\left(e^{z}\right) = \left(\frac{-4 \pm \sqrt{12}}{2}\right) \tag{E}$$

Taking the natural logarithm for both sides of equation (E), yields:

$$z = \ln\left(\frac{-4 \pm \sqrt{12}}{2}\right) \tag{F}$$

Only one solution can be accepted due to the basic definition of the logarithmic function, it will take the following form:

$$z = \ln\left(\frac{-4+\sqrt{12}}{2}\right) \tag{G}$$

Example

Prove that $\overline{e^z} = e^{\bar{z}}$.

Solution

Let us start with the complex-valued complex function e^z as follows:

$$e^z = u + iv = e^{x+iy} \tag{A}$$

The term e^{x+iy} can be re-written as follows:

$$e^{x+iy} = e^x \times e^{iy} \tag{B}$$

Making use of equations (A) and (B), one can get the following:

$$e^x \times e^{iy} = e^x \times (\cos y + i \sin y) \tag{C}$$

Now let us do the following mathematical manipulation:

$$\overline{e^z} = \overline{e^x \times e^{iy}} = e^x \times \left(\overline{\cos y + i \sin y}\right) \tag{D}$$

Now then equation (D) will be:

$$\overline{e^z} = e^x \times (\cos y - i \sin y) \tag{E}$$

Equation (E) is now takes the following form:

$$\overline{e^z} = e^x \times e^{-iy} = e^{x-iy} = e^{\bar{z}} \tag{F}$$

Equation (F), represents the required solution.

Example

Is $\overline{\ln z} = \ln \bar{z}$?

Solution

Let us start the solution by writing the following complex-valued function:

$$w = \ln z \tag{A}$$

Equation (A) can be re-written as follows:

$$z = e^{w} = e^{u+iv} \tag{B}$$

Now

$$\bar{z} = e^{\overline{u+iv}} = e^{u-iv} \tag{C}$$

Equation (C) can be re-cast as follows:

$$\ln \bar{z} = u - iv \tag{D}$$

Now let us dealing with the left hand-side of $\overline{\ln z} = \ln \bar{z}$

$$\overline{\ln z} = \bar{w} = \overline{u + iv}$$

$$\overline{\ln z} = u - iv \tag{E}$$

From equations (D) and (E), we get $\overline{\ln z} = \ln \bar{z}$

Example

Is $\overline{\sin z} = \sin \bar{z}$?

Solution

We will leave the solution for the reader.

Example

Show that $\tan z = \dfrac{\sin 2x + i \sinh 2y}{\cos 2x + \cosh 2y}$

Solution

$$\text{L.H.S.} = \tan z = \frac{\sin z}{\cos z}$$

$$= \frac{\sin(x+iy)}{\cos(x+iy)} = \frac{\sin x \cos iy + \cos x \sin iy}{\cos x \cos iy - \sin x \sin iy} \tag{A}$$

Note

$$\begin{aligned} \sin iy &= i \sinh y \\ \cos iy &= \cosh y \end{aligned} \tag{B}$$

Make use of equation (B) into (A), leads to:

$$\text{L.H.S.} = \frac{\sin x \cosh y + i \cos x \sinh y}{\cos x \cosh y - i \sin x \sinh y} \tag{C}$$

Let us do some mathematical manipulation on equation (C) as follows:

$$\text{L.H.S.} = \frac{\sin x \cosh y + i \cos x \sinh y}{\cos x \cosh y - i \sin x \sinh y} \times \frac{\cos x \cosh y + i \sin x \sinh y}{\cos x \cosh y + i \sin x \sinh y} \tag{D}$$

Simplify equation (D), leads to:

$$\text{L.H.S.} = \frac{\sin x \cos x \left(\cosh^2 y - \sinh^2 y\right) + i \sinh y \cosh y \left(\cos^2 x + \sin^2 x\right)}{\left(\cos^2 x \cosh^2 y + \sin^2 x \sinh^2 y\right)} \tag{E}$$

Note that:

$$\begin{aligned} \cosh^2 y - \sinh^2 y &= 1 \\ \cos^2 x + \sin^2 x &= 1 \end{aligned} \tag{F}$$

Making use of equation (F) into (E), the later takes the following simplifying form:

$$\text{L.H.S.} = \frac{\overbrace{\sin x \cos x}^{=\frac{1}{2}\sin 2x} + i \overbrace{\sinh y \cosh y}^{=\frac{1}{2}\sinh 2y}}{\left(\cos^2 x \cosh^2 y + \sin^2 x \sinh^2 y\right)} \tag{G-1}$$

$$\textbf{L.H.S.} = \frac{\frac{1}{2}\sin 2x + i\frac{1}{2}\sinh 2y}{\left(\cos^2 x\cosh^2 y + \sin^2 x\sinh^2 y\right)} \quad \text{(G-2)}$$

Note again

$$\sinh^2 y = \frac{1}{2}(\cosh 2y - 1)$$

$$\cosh^2 y = \frac{1}{2}(\cosh 2y + 1)$$

$$\cos 2x = 2\cos^2 x - 1 \quad \text{(G-3)}$$

Now the left-hand side after simplification takes the form:

$$\textbf{L.H.S.} = \frac{\sin 2x + i\sinh 2y}{(\cos 2x + \cosh y)} = \textbf{R.H.S.} \quad \text{(H)}$$

Exercises

1. Find $f(1+i)$ for the following functions:

(a) $f(z) = z + z^{-2} + 5$

(b) $f(z) = \dfrac{1}{z^2 + 1}$

(c) $f(z) = f(x+iy) = x + y + i\left(x^3 y - y^2\right)$

(d) $f(z) = z^2 + 4z\bar{z} - 5\,\text{Re}(z) + \text{Im}(z)$

2. Let $f(z) = z^{21} - 5z^7 + 9z^5$. Use polar coordinates to find:

(a) $f(-1+i)$

(b) $f\left(1+i\sqrt{3}\right)$

3. Express the following functions in the form $u(x,y)+iv(x,y)$:

(a) $f(z)=z^3$

(b) $f(z)=\bar{z}^{-2}+(2-3i)z$

4. Express the following functions in the polar form $u(r,\theta)+iv(r,\theta)$:

(a) $f(z)=z^5+\bar{z}^{-5}$

(b) $f(z)=z^5+\bar{z}^{-3}$

5. Let $f(z)=f(x+iy)=e^x\cos y+ie^x\sin y$. Find:

(a) $f(0)$

(b) $f(i\pi)$

6. Let $w=f(z)=(3+4i)z-2+i$

(a) Find the image of the disk $|z-1|<1$

(b) Find the image of the line $x=t, y=1-2t$, for $-\infty<t<\infty$

(c) Find the image of the half-plane $\text{Im}(z)>1$

7. Let $w=(2+i)z-2i$. Find the triangle onto which the triangle with vertices $z_1=-2+i$, $z_2=-2+2i$ **and** $z_3=2+i$ is mapped.

8. Let $S(z)=Kz, K>0$ is a positive real constant. Show that the equation $|S(z_1)-S(z_2)|=K|z_1-z_2|$ holds.

9. Find the linear transformation $w=f(z)$ that satisfy the following conditions:

(a) The points $z_1 = 2,\ z_2 = -3i$ maps onto $w_1 = 1 + i,\ w_2 = 1$

(b) The circle $|z| = 1$ maps onto the circle $|w - 3 + 2i| = 5$ and $f(-i) = 3 + 3i$

CHAPTER 3

Methods of Evaluating Complex Integrals

Abstract: Complex integration is an important topic in the field of complex analysis. There exist some related criteria between the integration in the real and complex functions. We will start this chapter, by introducing a short review for the line integration in the real domain then followed it by the subject of the present chapter. Theorems, corollaries and other facts will be introduced. After that we now turn to the problem of integrating complex functions. The theory you will learn is elegant, powerful, and a useful tool for physicists and engineers.

Keywords: Line integral, complex integral, independence of the path, annulus theorem, Cauchy integral theorem, generalized Cauchy integral theorem.

3.1. REVISION ON LINE INTEGRAL IN REAL DOMAIN

Before dealing with complex integrals, we found and due to the connectivity in so many ideas that there exist corresponding between integrals in real domain and that one in complex plane. To do so, assume that we have any function $M(x,y)$ and C as any smooth curve, *i.e.*, x & y can be represented parametrically as:

$$x=\Phi(t) \quad \& \quad y=\Psi(t), \quad \alpha\le t\le\beta \tag{1}$$

Then, the line integral for the function $M(x,y)$ over the path C is given by:

$$\int_C M(x,y)\,dx \qquad \text{or} \qquad \int_C N(x,y)\,dy \tag{2}$$

3.2. EXISTENCE OF LINE INTEGRAL

Suppose that:

1. C is a smooth curve expressed by:
   $x=\Phi(t)$ & $y=\Psi(t), \quad \alpha\le t\le\beta$

2. The function $M(x,y)$ is continuous on C.

Then, the integral of $M(x,y)$ along C with respect to x exists and is given by:

$$\int_C M(x,y)\,dx=\int_\alpha^\beta M(\Phi(t),\Psi(t))\,\Phi'(t)\,dt \tag{3}$$

S. G. Ahmed

or

$$\int_C M(x,y)\,dy = \int_{\alpha^*}^{\beta^*} M(\Phi(t),\Psi(t))\Psi^{/}(t)\,dt \tag{4}$$

The above integrals in terms of Cartesian coordinate system:

$$\int_C M(x,y)\,dx = \int_a^c M(x,f(x))\,dx \tag{5}$$

or

$$\int_C M(x,y)\,dy = \int_b^d M(g(y),y)\,dy \tag{6}$$

3.3. THE COMPLEX INTEGRAL

Let C be a smooth curve with $x=\Phi(t)$ & $y=\Psi(t)$, $\alpha \le t \le \beta$ and can be expressed as follows **[16]**:

$$C:\quad z(t)=x(t)+iy(t),\ \alpha \le t \le \beta \tag{7}$$

Suppose that $w=f(z)$ is a single-valued function at each point on C, then the line integral of $w=f(z)$ along C exists and takes the following form:

$$\begin{aligned}\int_C f(z)\,dz &= \int_C f(x+iy)\,d(x+iy)\\ &= \int_C f(x+iy)\,d(x+iy)\\ &= \int_C (u+iv)\,d(x+iy)\\ &= \underbrace{\int_C udx - vdy}_{\text{Real part of the integration}} + i\underbrace{\int_C vdx + udy}_{\text{Imginary part of the integration}}\end{aligned} \tag{8}$$

Example

Evaluate the line integral for:

$$f(z) = x$$

Along the contour shown in Fig. (**1**)

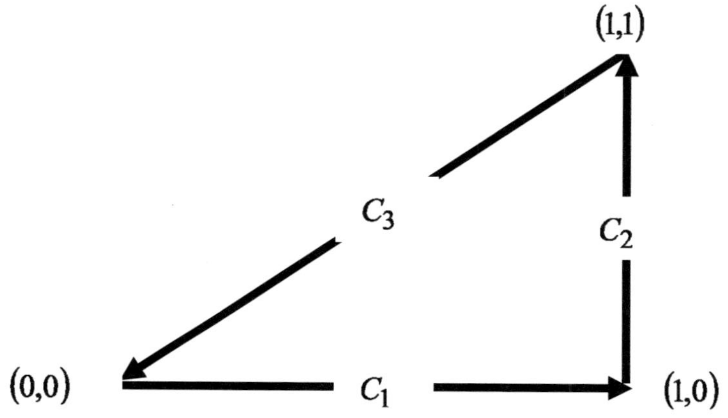

Figure 1: Problem configuration.

Solution

The path of integration can be represented in terms of Cartesian coordinates system as follows:

$$C_1: \;\; y = 0, \quad 0 \leq x \leq 1 \Rightarrow dy = 0$$

$$C_2: \;\; x = 1, \quad 0 \leq y \leq 1 \Rightarrow dx = 0$$

$$C_3: \;\; y = x, \quad (1,1) \rightarrow (0,0) \Rightarrow dx = dy$$

Now, let us evaluate the integral over the three paths shown before, as follows:

$$\int_C f(z)\,dz = \int_{C_1+C_2+C_3} f(z)\,dz = \int_{C_1} f(z)\,dz + \int_{C_2} f(z)\,dz + \int_{C_3} f(z)\,dz$$

$$= \int_{C_1} f(x+iy)\,d(x+iy) + \int_{C_2} f(x+iy)\,d(x+iy) + \int_{C_3} f(x+iy)\,d(x+iy)$$

$$= \int_{C_1} f(x+iy)d(x+iy) + \int_{C_2} f(x+iy)d(x+iy) + \int_{C_3} f(x+iy)d(x+iy)$$

$$= \int_{C_1} x\left(dx + i\underbrace{dy}_{\text{zero}}\right) + \int_{C_2} x\left(\underbrace{dx}_{\text{zero}} + idy\right) + \int_{C_3} x\left(dx + i\underbrace{dy}_{=dx}\right)$$

$$= \int_{x=0}^{x=1} xdx + i\int_{x=0}^{x=1} xdy + (1+i)\int_{x=0}^{x=1} xdx = \frac{i}{2}$$

Example

Evaluate the following integral $\int_C \frac{dz}{z-i}$, C is the circle $|z-i|=1$

Solution

The equation of the path can be re-written in terms of polar coordinates as follows see Fig. (**2**):

$$z = re^{it}, \quad r = 1 \;\; \& \;\; 0 \le t \le 2\pi$$

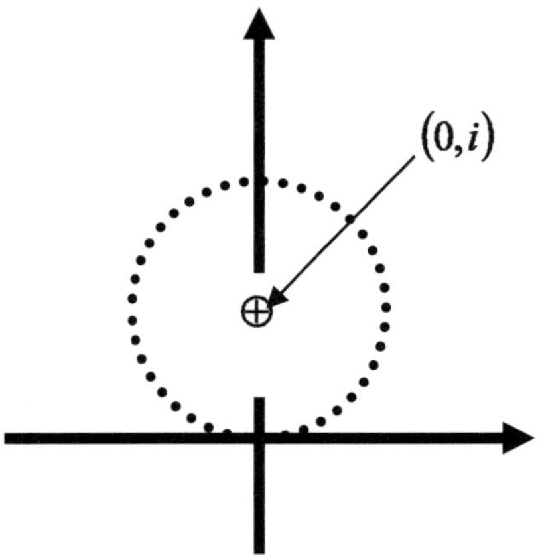

Figure 2: Problem configuration.

$$\int_C \frac{dz}{z-z_0} = \int_0^{2\pi} \frac{rie^{it}}{re^{it}} dt = \int_0^{2\pi} idt = 2\pi i$$

what we want from this example, is that the integral on a circle of a center at the point of singularity equals $2\pi i$, and this result will be used after that as rule.

3.4. CAUCHY THEOREM OF INTEGRATION

Before, we learned how to evaluate the line integral, without any restriction on the integrand – the function inside the integral sign –. In the present section, we will focus the lights on how to evaluate the integral containing analytical functions, which require more attention **[17]**.

Suppose that:

1. $f(z)$ is analytic function on a simply connected region R,

2. C is a closed path lying entirely in R.

Then

$$\oint_C f(z)\,dz = 0 \tag{9}$$

Remark

The converse of Cauchy's theorem does not hold, *i.e.*, the following statement false "If $\oint f(z)\,dz = 0$ $\forall$ closed path C in a simply connected region R, then $f(z)$ is analytric in R see Fig. **3**.

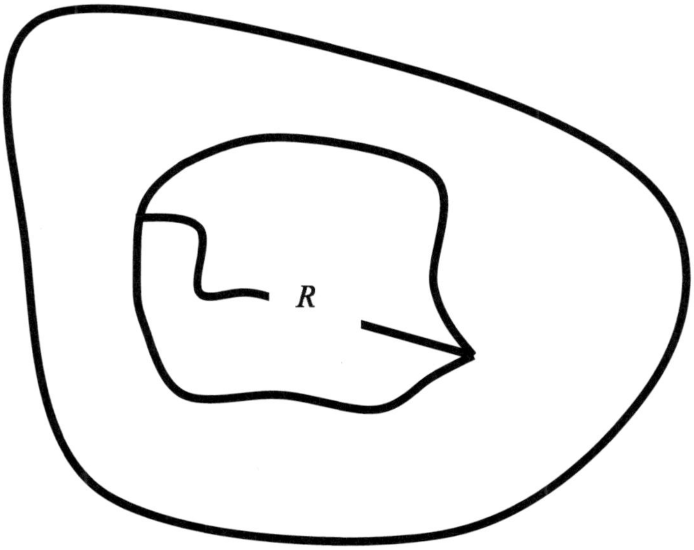

Figure 3: Independence of the path.

3.5. INDEPENDENCE OF THE PATH

Suppose that:

1. R is a simply connected region R ,

2. z_1 and z_2 are two points in R ,

3. $f(z)$ is analytic function on the simply connected region R . Then, the value of $\int_{z_1}^{z_2} f(z)\,dz$ is the same along any path joining the two points, provided that the path lies entirely within R , see Fig. (**3**) **[18]**.

3.6. ANNULUS OR ANNULAR REGION

The annulus is defined by the region between two simple closed paths, as shown in Fig. (**4**).

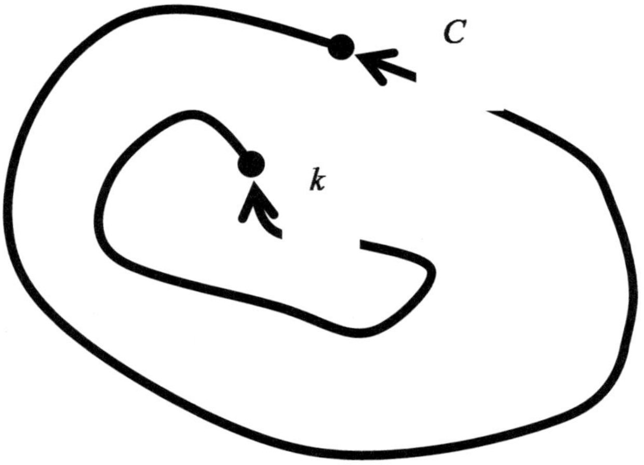

Figure 4: Annulus between two simple closed paths.

3.6.1. Annulus Theorem

Suppose that $f(z)$ is analytic on the closed annulus region determined by two simply closed paths C and k , respectively then **[19]**:

$$\int_C f(z)dz = \int_k f(z)dz \tag{10}$$

such that C and k have the same positive orientation.

Example

Evaluate the integral $\int_C \frac{dz}{z-i}$ where:

C is the path shown in Fig. (**5**).

Solution

First, we can say that there exists a point of singularity at $z = i$

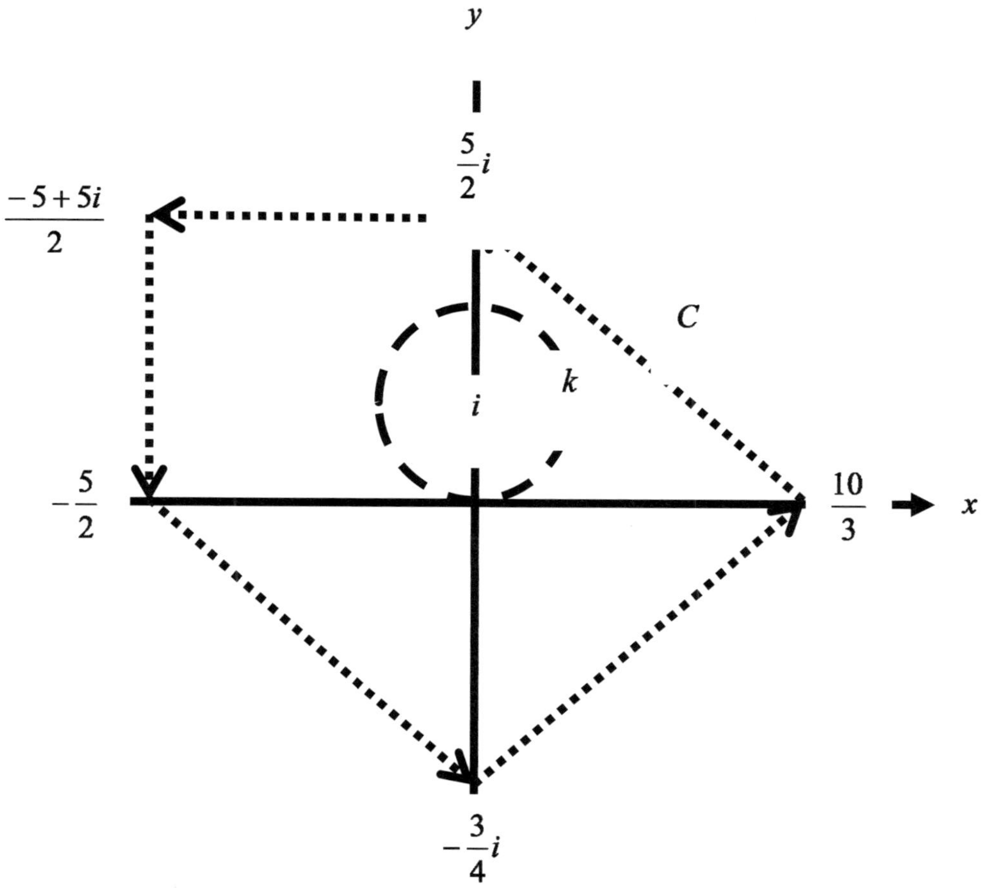

Figure 5: Domain problem

Take a circle with center at the point of singularity at $z = i$ and very small radius length inside the domain of the problem and has the same positive orientation. This means that the function $f(z) = \dfrac{1}{z-i}$ will be analytic in the region bounded by the main path C and the interior path k, apply then the annulus theorem, one can get:

$$\int_C f(z)dz = \int_k f(z)dz = \int_k \frac{1}{z-i}dz = 2\pi i$$

Example

Evaluate the following integral $\int_C \left(\dfrac{4}{z+1} + \dfrac{3}{z+2i}\right)dz$, where C is the circle: $|z| = 4$

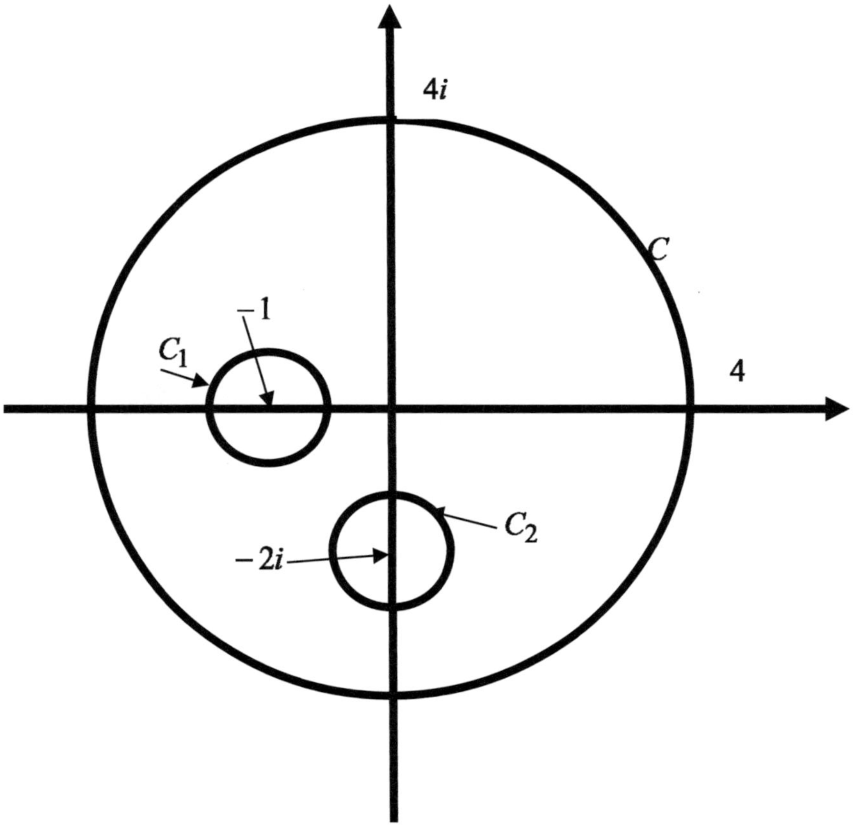

Figure 6: Domain problem.

Solution

The first step is to determine the points of singularities – if any -, the integrand is not analytic at $z=-1$ and $z=-2i$, then next step is to take circles with centers at the point of singularities, have very small radii and have the same orientation of the main path, see Fig. (**6**).

Now

$$\int_C f(z)dz = \int_{C_1} f(z)dz + \int_{C2} f(z)dz$$

Then

$$\int_C f(z)dz = \int_{C_1}\left(\frac{4}{z+1}+\frac{3}{z+2i}\right)dz + \int_{C_2}\left(\frac{4}{z+1}+\frac{3}{z+2i}\right)dz$$

$$\int_C f(z)dz = 4\underbrace{\int_{C_1}\left(\frac{1}{z+1}\right)dz}_{I_1} + 3\underbrace{\int_{C_1}\left(\frac{1}{z+2i}\right)dz}_{I_2}$$

$$+4\underbrace{\int_{C_2}\left(\frac{1}{z+1}\right)dz}_{I_3} + 3\underbrace{\int_{C_2}\left(\frac{1}{z+2i}\right)dz}_{I_4}$$

$\underbrace{\int_{C_1}\left(\frac{1}{z+1}\right)dz}_{I_1} = (2\pi i)$ because the function is not analytic at $z=-1$ over C_1

$\underbrace{\int_{C_1}\left(\frac{1}{z+2i}\right)dz}_{I_2} = 0$ because the function is analytic overC_1

$\underbrace{\int_{C_2}\left(\frac{1}{z+1}\right)dz}_{I_3} = 0$ because the function is analytic overC_2

$\underbrace{\int_{C_2}\left(\frac{1}{z+2i}\right)dz}_{I_4} = (2\pi i)$ because the function is not analytic at $z = -2i$ over C_2

Then

$$\int_C f(z)dz = 4(2\pi i) + 3(0) + 4(0) + 3(2\pi i) = 14\pi i$$

3.7. MORERA'S THEOREM

Suppose that **[19]**:

1. The function $f(z)$ is analytic **<u>on</u>** a positively oriented simple closed and interior path C

2. The point z_0 is any point interior the path C

Then

$$f(z_0) = \frac{1}{2\pi i}\int_C \frac{f(z)}{(z-z_0)}dz \tag{11}$$

Note: The word on in the above theorem means that the function under integration may have singular point in R or interior of the path C.

Remark

1. It is important to distinguish between the function $f(z)$ which is analytic interior C and the function $\frac{f(z)}{(z-z_0)}$ which has a singularity interior C at the point $(z = z_0)$.

2. By hypothesis it is assumed that the path C is positive direction of orientation but if C is of negative orientation then, the left-hand side of the above theorem will be $\left[-f(z_0)\right]$.

Next, the main steps to evaluate the integral using Morera's theorem are summarized as follow:

1. We have an integral containing the integrand $f(z)$.

2. Choose $f(z)$ such that to be analytic everywhere and even at the point $z = z_0$ in the integral.

3. Evaluate $f(z)$ at the point of singularity $z = z_0$

4. Finally, the value of the integration $\int_C \frac{f(z)}{(z-z_0)} dz$ take the value $(2\pi i) f(z_0)$

Hint: It is important to mention how can we choose the function $f(z)$? This will be seen clearly through the following examples.

Example

Evaluate the following integral:

$I = \int_C \frac{z^2}{(z-i)} dz$, $C : |z| = 2$ The path is positively oriented.

Solution

As it seen from the integral given that the integrand has a singularity at the point $z = i$, so if we remove this part from the integrand, one can get a new function, that is $f(z)$, This new function is analytic entirely the path C including the point of singularity $z = i$.

Now then,

$$f(z_0) = f(i) = (i)^2 = -1$$

Therefore;

$$I = \int_C \frac{z^2}{(z-i)} dz = f(z_0) = -1$$

$$= \frac{1}{2\pi i} \int_C \frac{f(z)}{(z-i)} dz$$

$$\Rightarrow$$

$$\int_C \frac{f(z)}{(z-i)} dz = -2\pi i$$

Example

Evaluate the following integral:

$$I=\int_C \frac{1}{z\left(z+\pi i\right)}dz\,,\ C:z=-3i+e^{it},\quad 0\le t\le 2\pi$$

The path is positively oriented.

Solution

As it seen from the integral given that the integrand has a singularity at the points $z=0\,\&\,z=-\pi i$, so if we remove part of this singularity from the integrand, one can get a new function $f\left(z\right)$ as follows:

$$f\left(z\right)=\left(\frac{1}{z+\pi i}\right)$$

Then we want to evaluate the following integral:

$$I=\int_C \frac{\left(\frac{1}{z+\pi i}\right)}{z}dz$$

with point of singularity at $z=0$

Now then,

$$f\left(z_0\right)_{z_0=0}=\left(\frac{1}{z_0+\pi i}\right)=\left(\frac{1}{\pi i}\right)$$

Therefore;

$$\frac{1}{2\pi i}\int_C \frac{\left(\frac{1}{z+\pi i}\right)}{z}dz=f\left(z_0\right)=\left(\frac{1}{\pi i}\right)$$

$$\int_C \frac{\left(\frac{1}{z+\pi i}\right)}{z} dz = \left(\frac{2\pi i}{\pi i}\right) = 2$$

$$\Rightarrow$$

$$I = \int_C \frac{\left(\frac{1}{z+\pi i}\right)}{z} dz = 2$$

3.8. GENERALIZED CAUCHY FOR INTEGRATION

Suppose that:

1. The function $f(z)$ is analytic over and inside a positive closed path C.
2. The point z_0 is any point interior C,

Then, for any integer $n = 0, 1, 2, \ldots$ the derivative $f^{(n)}(z_0)$ exists and is given by **[20]**:

$$f^{(n)}(z_0) = \frac{n!}{2\pi i} \int_C \frac{f(z)}{(z - z_0)^{n+1}} dz$$

Corollary

Suppose that $f(z)$ is analytic in a region R. Then $f(z)$ has derivatives of all orders each of which is analytic in R and is given by the formula of generalized Cauchy integral formula.

Example

Evaluate $I = \int_C \frac{z^3 + e^z}{(z + \pi i)^3} dz,\ C: z = 7e^{it},\ 0 \le t \le 2\pi$

The path is a circle with center at the origin and of radius equals 7 units.

Solution

First, we will choose $f(z)$ in such a way that it will be analytic over and interior the path C.

$$f(z) = z^3 + e^z$$

Re-write the generalized Cauchy integral formula

$$f^{(n)}(z_0) = \frac{n!}{2\pi i} \int_C \frac{f(z)}{(z - z_0)^{n+1}} dz$$

Now

$$z_0 = -\pi i$$

By comparing the integral under-hand and the generalized Cauchy integral, one can get the following:

$$n + 1 = 3 \Rightarrow n = 2$$

Evaluation the derivatives of the new function:

$$f'(z) = 3z^2 + e^z$$

$$f''(z) = 6z + e^z$$

Evaluation the 2nd derivative at the point of singularity, leads to:

$$f''(-\pi i) = 6(-\pi i) + e^{(-\pi i)} = -6\pi i + e^{(-\pi i)}$$

Now we reached to the last step, is to go back to the generalized Cauchy integral formula to find the value of the given integral as follows:

$$f^{(2)}(z_0) = \frac{2!}{2\pi i} \int_C \frac{f(z)}{(z - z_0)^{n+1}} dz$$

$$\Rightarrow$$

$$\int_C \frac{f(z)}{(z-z_0)^{n+1}}dz = \frac{2\pi i!}{2!}f^{(2)}(z_0) = \frac{2\pi i!}{2!}\left\{-6\pi i + e^{(-\pi i)}\right\} = \left\{6\pi^2 - \pi i e^{(-\pi i)}\right\}$$

Example

$$\text{Evaluate } I = \int_C \frac{dz}{z^3(z-2i)^2},\ C: |z-3| = 2$$

The path is positively oriented and it is a circle with center at $(3,0)$.

Solution

First, we will choose $f(z)$ in such a way that it will be analytic over and interior the path C.

$$f(z) = \frac{1}{z^3}$$

Re-write the generalized Cauchy integral formula

$$f^{(n)}(z_0) = \frac{n!}{2\pi i}\int_C \frac{f(z)}{(z-z_0)^{n+1}}dz$$

Now

$$z_0 = 2$$

By comparing the integral under-hand and the generalized Cauchy integral, one can get the following:

$$n+1 = 2 \Rightarrow n = 1$$

So we need to find the first derivative only of the new function:

$$f'(z) = \frac{-3z^2}{z^2} = -3\frac{1}{z^4}$$

The 1st derivative at the point of singularity leads to:

$$f^{/}(2)=\frac{-3}{(2)^4}=\frac{-3}{16}$$

Now we reached to the last step, is to go back to the generalized Cauchy integral formula to find the value of the given integral as follows:

$$f^{(1)}(z_0)=\frac{2!}{2\pi i}\int_C \frac{f(z)}{(z-z_0)^2}dz$$

$$\Rightarrow$$

$$\int_C \frac{f(z)}{(z-z_0)^{n+1}}dz=\frac{2\pi i!}{1!}f^{(1)}(z_0)$$

$$=\frac{2\pi i!}{1!}\left\{\frac{-3}{16}\right\}$$

$$=\left\{\frac{-3}{8}\pi i\right\}$$

Example

Evaluate $\oint_C (z^2-z+2)dz,\ C:|z|=1$

Solution

The integrand is a polynomial of order 2 and has no any points of singularity therefore it is analytic over the region R which is a circle with center at the origin and of unit radius. From Cauchy theorem of integration, the integral equals zero, *i.e.*,

$$\oint_C (z^2-z+2)dz,=0$$

Example

Evaluate $\oint_C \left(\frac{z}{2z-5}\right)dz,\ C:|z|=2$

Solution

The region of integration is a circle with center at the origin and of radius 2 units.

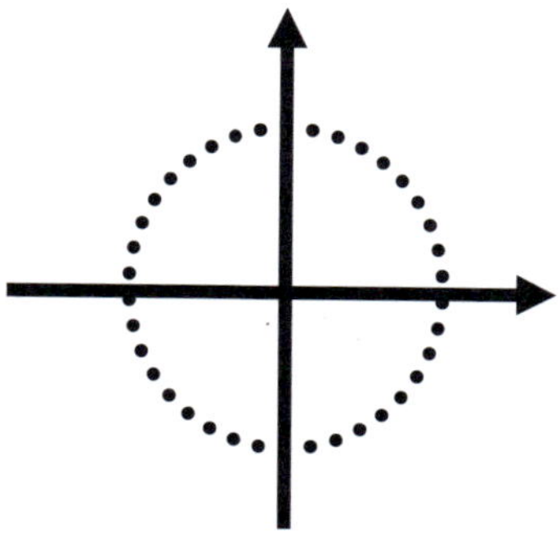

Figure 7: Domain problem.

The integrand has a singularity at the point $z_0 = \frac{5}{2} = 2.5$, *i.e.*, the point of singularity is outside the domain of the problem. Therefore, due to Cauchy integral theorem, the value of the given integration will be zero.

Example

Evaluate $\oint_C \left(\frac{z}{2z-5} \right) dz, \; C: |z-3| = 2$

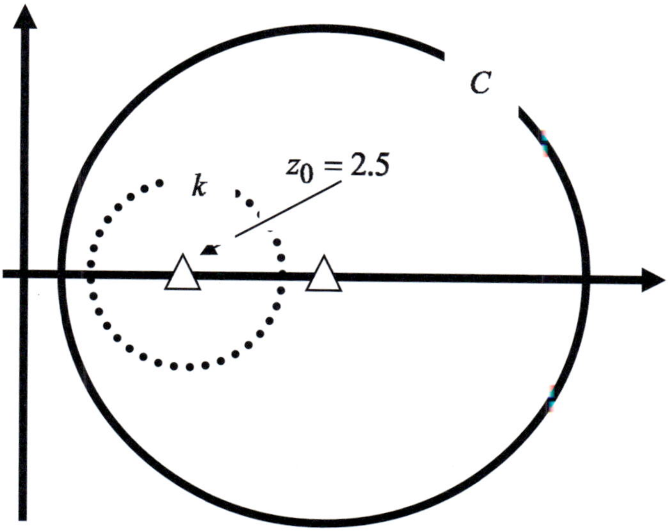

Figure 8: Domain problem.

Solution

The region of integration is a circle with center at the point $(3,0)$ and of radius 2 units.

The integrand has a singularity at the point $z_0 = \frac{5}{2} = 2.5$, *i.e.*, the point of singularity is inside the domain of the problem. Therefore, to apply Cauchy integral theorem, one can isolate the point of singularity by choosing a new function $f(z)$ to be analytic inside the region of integration. Let us choose this function as follows:

$$f(z) = z$$

Now

$$\oint_k \left(\frac{z}{2z-5} \right) dz = 2\pi i \left[f(z_0) \right] = 2\pi i \left[\frac{5}{2} \right] = 5\pi i$$

Example

$$\text{Evaluate} \oint_C \left(\frac{z+1}{z^2(z-2)} \right) dz, \; C: |z| = 1$$

Solution

Firstly, let us do some mathematical manipulation for the integrand using partial fractions as follows:

$$\frac{z+1}{z^2(z-2)} = \frac{A}{(z-2)} + \frac{Bz+C}{z^2}$$

$$z+1 = Az^2 + (Bz+C)(z-2)$$

$$z+1 = z^2(A+B) + z(C-2B) - 2C$$

Equating the coefficients of the same powers of z, leads to:

$$A = \frac{3}{4}$$

$$B = -\frac{3}{4}$$

$$C = -\frac{1}{2}$$

Now the integral takes the following new form:

$$\oint_C \left(\frac{z+1}{z^2(z-2)} \right) dz = \oint_C \left(\frac{A}{(z-2)} + \frac{Bz}{z^2} + \frac{C}{z^2} \right) dz$$

$$\oint_C \left(\frac{z+1}{z^2(z-2)} \right) dz = \oint_C \left(\frac{\frac{3}{4}}{(z-2)} + \frac{-\frac{3}{4}}{z} + \frac{-\frac{1}{2}}{z^2} \right) dz$$

$$\oint_C \left(\frac{z+1}{z^2(z-2)} \right) dz = \frac{3}{4} \oint_C \left(\frac{dz}{(z-2)} \right) - \frac{3}{4} \oint_C \left(\frac{dz}{z} \right) - \frac{1}{2} \oint_C \left(\frac{dz}{z^2} \right)$$

Next step is to evaluate the three integrals obtained before over the defined in the example;

The First Integral

The integrand is analytic over and interior the path C, then the value of the integral will be:

$$I_1 = \frac{3}{4} \oint_C \frac{dz}{(z-2)} = 0$$

The Second Integral

The integrand has singularity interior the path C, then the value of the integral will be:

$$I_2 = -\frac{3}{4}\oint_C \frac{dz}{z} = 2\pi i$$

The Third Integral

The integrand has singularity interior the path C, then the value of the integral will be:

$$I_3 = -\frac{1}{2}\oint_C \frac{dz}{z^2} == 2\pi i$$

Finally; the value of the original integral will be:

$$I = \frac{3}{4}I_1 - \frac{3}{4}I_2 - \frac{1}{2}I_3 = -\left(\frac{5\pi i}{2}\right)$$

Integrals of the Form

$$I = \int_0^{2\pi} \frac{dt}{a + b\sin t}$$

or

$$I = \int_0^{2\pi} \frac{dt}{a + b\cos t}$$

Such that $|a| > |b|$, in all these integrals, one can use the following relations:

$$\cos t = \frac{1}{2}\left(z + \frac{1}{z}\right)$$

$$\sin t = \frac{1}{2i}\left(z - \frac{1}{z}\right)$$

$$dt = -\frac{i}{z}dz$$

Example

Prove that:

$$I = \int_0^{2\pi} (\cos\Theta)^{2n}\, d\Theta = \frac{(2n)!\pi}{2^{2n-1}(n!)^2},\ n = 0,1,2,\ldots.,\ \text{with path}\, C:|\,z\,| = 1$$

Solution

Let us start solution by dealing with the left-hand side, and make use of:

$$\cos\Theta = \frac{1}{2}\left(z + \frac{1}{z}\right) \tag{A-1}$$

with

$$d\Theta = -\frac{i}{z}dz \tag{A-2}$$

Then, the left-hand side will be:

$$\text{L.H.S.} = \frac{1}{2^{2n}}\int_C \left(z + \frac{1}{z}\right)^{2n}\left(\frac{-i}{z}\right)dz \tag{A-3}$$

Let us now deals with the brackets inside the integrand as follows:

$$\left(z + \frac{1}{z}\right)^{2n} = \left(z + z^{-1}\right)^{2n}$$

$$= (z)^n + (2n)(z)^{n-1}(z)^{-1} + \frac{(2n)(2n-1)}{2!}(z)^{2n-2}(z)^{-2} + \ldots..$$

Simplify, we get:

$$\left(z + z^{-1}\right)^{2n} = (z)^n + (2n)(z)^{n-2} + \frac{(2n)(2n-1)}{2!}(z)^{2n-4} + \ldots.. \tag{A-4}$$

The remaining steps are left to the reader.

Definition

Let us turn to another method for evaluation the complex definite integral by assuming that the complex function will take the following form:

$f(t)=u(t)+iv(t)$, where *u* and *v* are real-valued functions of the real variable *t*, for $a \le t \le b$.

Then:

$$\int_a^b f(t)\,dt=\int_a^b u(t)\,dt+i\int_a^b v(t)\,dt \tag{1}$$

By carrying out the integration by a normal manner, leads to:

$$\int_a^b f(t)\,dt=\left[U(t)+iV(t)\right]_a^b=\left[\left(U(b)-U(a)\right)+i\left(V(b)-V(a)\right)\right] \tag{2}$$

Example

Evaluate the following definite integral $\int_0^1 (t-i)^3\,dt$

Solution

Before evaluating the given integral, let us re-write the integral as follows:

$$f(t)=(t-i)^3=t^3-3t+i\left(-3t^2+1\right).$$

Then, one obtain both $u(t)$ and $v(t)$ as follows, respectively:

$$u(t)=t^3-3t$$

$$v(t)=i\left(-3t^2+1\right)$$

Let us now evaluate the integrals as follow:

$$\int_0^1 u(t)\,dt=\int_0^1\left(t^3-3t\right)dt=-\frac{5}{4}$$

$$\int_0^1 v(t)\,dt = \int_0^1 \left(-3t^2+1\right) dt = 0$$

Hence, one can get the following result:

$$\int_0^1 (t-i)^3\,dt = \int_0^1 u(t)\,dt + i\int_0^1 v(t)\,dt = -\frac{5}{4}$$

Example

Evaluate $I = \int_{1-2i}^{3+i} (2z+3)\,dz$ along $x = 2t+1,\ y = 4t^2 - t - 2,\ 0 \le t \le 1$

Solution

Starting the solution by re-writing the given integral and replace the variable z by the corresponding x and y variables, leads to:

$$I = \int_{1-2i}^{3+i} (2x+2i+3)\,d(x+iy)$$

$$I = \int_{1-2i}^{3+i} \left[(2x+3)+2iy\right](dx+idy)$$

$$I = \int_{1-2i}^{3+i} \left[(2x+3)dx - 2ydy\right] + i\int_{1-2i}^{3+i} \left[(2x+3)dy + 2ydx\right]$$

Next step is to substitute the variables x and y by the parametric one, we leave the last few steps of calculation as a practice for the reader.

Some Properties

Let $f(t) = u(t) + iv(t)$ and $g(t) = p(t) + iq(t)$ be continuous on $a \le t \le b$.

Then,

$$\int_a^b (f(t)+g(t))\,dt = \int_a^b f(t)\,dt + \int_a^b g(t)\,dt \qquad (3)$$

Dividing the interval $a \le t \le b$ into two sub-intervals have no effect on the final result of the integral:

$$\int_a^b f(t)\,dt = \int_a^c f(t)\,dt + \int_c^b f(t)\,dt \tag{4}$$

If the integrand is multiplied by a complex constant $(c + id)$, then:

$$\int_a^b (c+id) f(t)\,dt = (c+id)\int_a^b f(t)\,dt \tag{5}$$

Also, the following property is valid:

$$\int_a^b f(t)\,dt = -\int_b^a f(t)\,dt \tag{6}$$

Product Property

$$\int_a^b f(t)g(t)\,dt = \int_{au}^b \left[u(t)p(t) - v(t)q(t)\right]dt + i\int_{au}^b \left[u(t)q(t) + v(t)p(t)\right]dt \tag{7}$$

3.9. CONTOURS AND CONTOUR INTEGRALS

In the previous sections, we learned how to evaluate an integral of the form $\int f(z)\,dz$ over a closed simple path called contour C **[21]**, the contour in its parametric form will be as follows:

$$C: z(t) = x(t) + iy(t), \quad a \le t \le b \tag{8}$$

In equation (8), $x(t)$ & $y(t)$ are continuous functions.

3.9.1. Closed Curve

A curve C with the property $z(b) = z(a)$ is called a closed curve.

3.9.2. Initial and Terminal points

The starting point of the contour C is called initial point and will denoted as $z(a)$ and the final point will then defined as terminal point and will denoted as $z(b)$.

3.9.3. Simple Closed Curve

Over the contour C, if there is only one point of intersection, then the contour is called simple closed curve.

3.9.4. Orientation

The orientation means movement over a contour C from the starting point to the terminal point or vise versa.

To understand the concept of orientation, refer to Fig. (**9**).

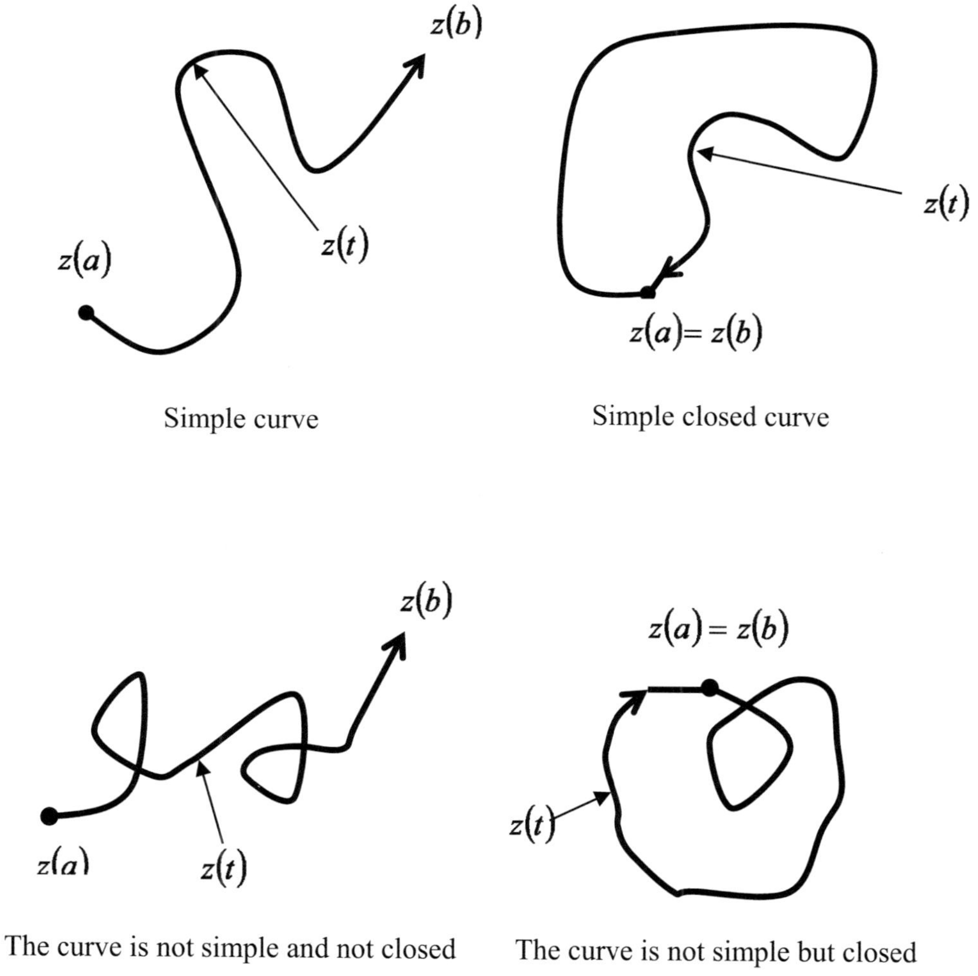

Figure 9: The terms simple and closed used to describe curves.

3.9.5. Differentiable complex function

A complex function in its parametric form $z(t) = x(t) + iy(t)$ is said to be differentiable on $[a, b]$ if both real part $x(t)$ and imaginary part $y(t)$, respectively are differentiable on the interval $a \le t \le b$. Before continuing let us remember what is called one-sided derivatives:

The derivatives on the right $x'(a^+)$ and on the left $x'(b^-)$ are defined by the limits:

$$x'(a^+) = \lim_{t \to a^+} \frac{x(t) - x(a)}{t - a}$$

$$x'(b^-) = \lim_{t \to b^-} \frac{x(t) - x(b)}{t - b}$$

Now the derivative z' is:

$$z'(t) = x'(t) + iy'(t) \text{ for } a \le t \le b.$$

3.9.6. Smooth Curve

If a curve C defined in its parametric form is continuous and non-zero on any closed interval, under these conditions, the curve C is called smooth **[22]**, see Fig. (**10**).

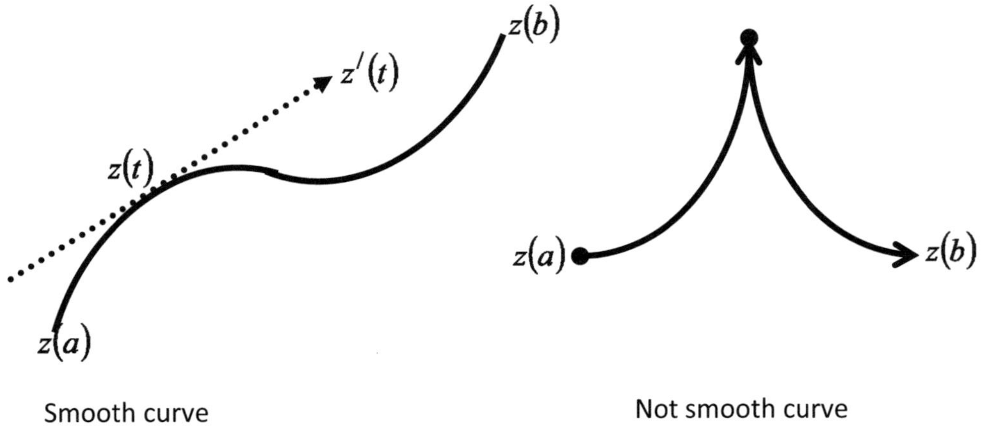

Figure 10: Smooth and not smooth curves.

For a smooth curve C, defined by its parametric form, the differential is defined as, $\Delta s = s(t)\Delta t$, in which the function $s(t)$ is continuous and its real and imaginary parts, respectively are differentiable over a closed interval. Therefore, the arc length of the smooth curve is given by:

$$L(C) = \int_a^b \sqrt{\left[x'(t)\right]^2 + \left[y'(t)\right]^2}\, dt = \int_a^b \left\|\left[z'(t)\right]\right\| dt \tag{9}$$

3.9.7. Opposite Curve

If one moves along the smooth curve from its terminal point to the initial one, the curve in this case is called the opposite curve, as in Fig. (**11**). Let the smooth curve in its parametric form is given by:

$$C : z_1(t) = x(t) + iy(t), \quad a \le t \le b$$

Then, the opposite curve can be defined as follows:

$$-C : z_2(t) = x(-t) + iy(-t), \quad -b \le t \le -a$$

Both C and its opposite one are shown in Fig. (**11**).

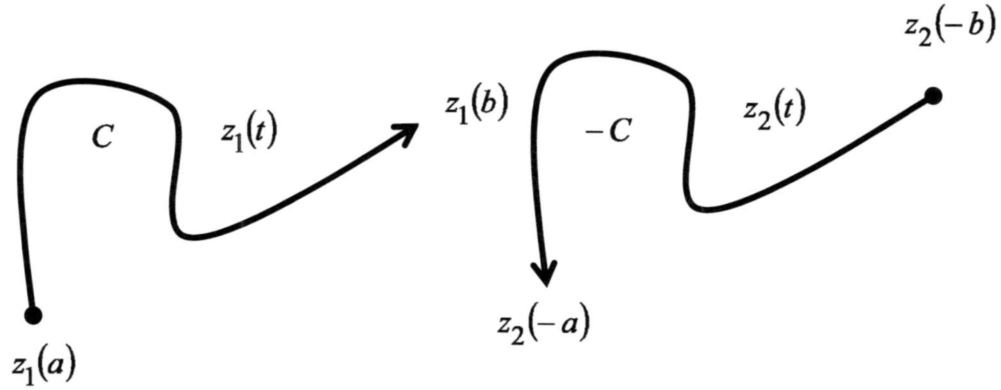

Figure 11: The curve C and its opposite –C.

Example

A straight line joining the two points $z_0(x_0, y_0)$ and $z_1(x_1, y_1)$. Find the equation of the curve C and its opposite in their parametric form.

Solution

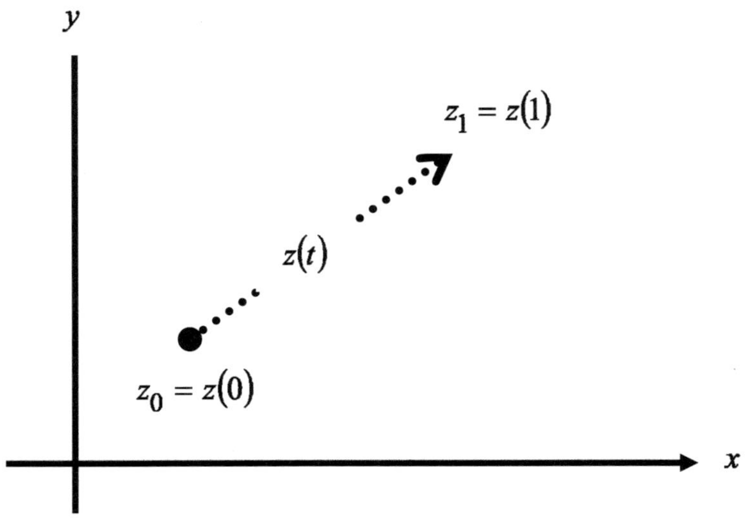

Figure 12: The straight-line segment C joining z_0 and z_1.

From Fig. (**12**), one can write vector form of the straight line as follows:

$$C : z(t) = z_0 + (z_1 - z_0)\ t \text{ for } 0 \le t \le 1 \tag{10-1}$$

or

$$C : z(t) = [x_0 + (x_1 - x_0)\ t] + i[y_0 + (y_1 - y_0)\ t] \text{ for } 0 \le t \le 1 \tag{10-2}$$

Clearly one parametrization for –C is

$$C : \gamma(t) = z_1 + (z_0 - z_1)\ t \text{ for } 0 \le t \le 1 \tag{10-3}$$

Example

For the broken straight line in Fig. (**13**). Find the equation of the curve C and its opposite in their parametric form.

Solution

Refer to Fig. (**13**) and assume that:

$$z_0 = -1 + i$$

and

$$z_1 = -1$$

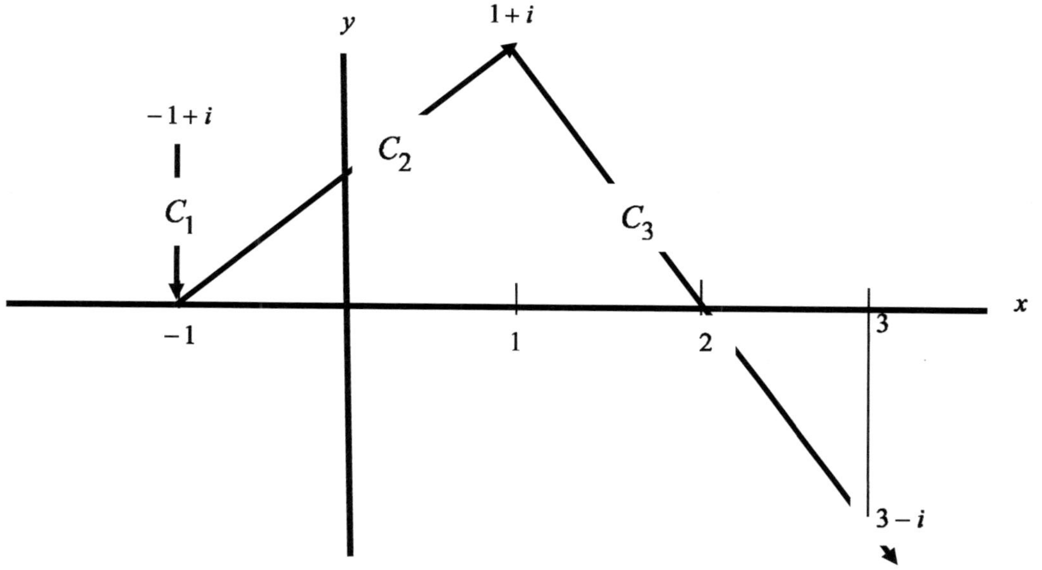

Figure 13: The path $C = C_1 + C_2 + C_3$ from $-1+i$ to $3-i$.

The parameterization of C_1 can take the following form:

$$C_1 : z_1(t) = z_0 + (z_1 - z_0)\ t = (-1+i) + t(-1-(-1+i)) \text{ for } 0 \le t \le 1$$

Simplification leads to:

$$C_1 : z_1(t) = -1 + i(1-t) \text{ for } 0 \le t \le 1$$

A similar procedure for C_2 and C_3, gives:

$$C_2 : z_2(t) = (-1+2t) + it \text{ for } 0 \le t \le 1$$

and

$$C_3 : z_3(t) = (1+2t) + i(1-2t) \ \text{for}\ 0 \le t \le 1$$

Theorem

If a complex function $f(z)$ is a continuous over a set containing the contour C, and let $f(z)$ be the parametric form of C for $a \le t \le b$.

Then:

$$\int_C f(z)\,dz = \int_a^b f(z(t))\,z'(t)\,dt$$

Example

Evaluate $\int_C f(z)\,dz$, over C, which is the line segment joining the two points $A = 0$ to $B = 2 + i\frac{\pi}{4}$

Solution

The first step is to write the parametric form of the line joining the given two points, leads to:

$$C : z(t) = \left(2 + i\frac{\pi}{4}\right) t \ \text{for}\ 0 \le t \le 1$$

Next step is to find the derivative of the parametric function, leads to:

$$z'(t) = \left(2 + i\frac{\pi}{4}\right)$$

Let us apply the previous theorem, as follows:

$$\int_C \exp(z)\,dz = \int_0^1 \exp\left(\left(2 + i\frac{\pi}{4}\right) t\right)\left(2 + i\frac{\pi}{4}\right) dt$$

$$\int_C \exp(z)\,dz = \left(2 + i\frac{\pi}{4}\right)\int_0^1 \exp(2\,t)\exp\left(i\frac{\pi}{4}\,t\right) dt$$

$$\int_C \exp(z)\,dz = \left(2+i\frac{\pi}{4}\right)\int_0^1 \exp(2\,t)\left(\cos\frac{\pi}{4}\,t + i\sin\frac{\pi}{4}\,t\right)dt$$

$$\int_C \exp(z)\,dz = \left(2+i\frac{\pi}{4}\right)\int_0^1 \exp(2\,t)\left(\cos\frac{\pi}{4}\,tt\right)dt + i\int_0^1 \exp(2\,t)\left(\sin\frac{\pi}{4}\,t\right)dt$$

Evaluation the integral in the right-hand side of the last equation can be occurred using the integration by parts. We will leave these evaluation for the reader.

Example

$$\text{Evaluate } \int_{C_1^+(2)} \left(\frac{1}{z-2}\right)dz\,.$$

Solution

Recall that $C_1^+(2)$ is the upper semi-circle with radius 1 centered at $x=0$ oriented in a positive direction *i.e.*, counterclockwise. Next, the parametrization of C is:

$$z(t) = 2 + e^{it},\ 0 \le t \le \pi$$

The next step is to apply the preceding theorem with $f(z) = \dfrac{1}{z-2}$ and the derivative of the parametrization $z^{/} = ie^{it}$

Therefore;

$$\int_C \left(\frac{1}{z-2}\right)dz = \int_0^{\pi} \frac{1}{\left(2+e^{it}\right)-2}\,ie^{it}\,dt = \int_0^{\pi} i\,dt = i\pi$$

3.10. COMPLEX DIFFERENTIAL

If $z(t) = x(t) + iy(t)$, then we have:

$$dz = z^{/}(t)\,dt = \left[x^{/}(t) + iy^{/}(t)\right]dt = dx + idy \tag{11}$$

where dx and dy are the differentials for $x(t)$ & $y(t)$, respectively. The expression dz is often called the ***complex differential*** **[23]**.

Alternative Method for Evaluating the Complex Integral

Suppose that:

Assume that we have a complex function $f(z)=u(z)+iv(z)$ over a contour C in its parametric form $z(t)=x(t)+iy(t)$, then:

$$\int_C f(z)dz=\int_a^b f(z(t))z'(t)dt=\int_a^b\left[u(z(t))+iv(z(t))\right]\left[x'(t)+iy'(t)\right]dt$$
$$=\int_a^b\left[u(z(t))x'(t)-v(z(t))y'(t)\right]dt+i\int_a^b\left[v(z(t))x'(t)+u(z(t))y'(t)\right]dt$$
$$=\int_a^b\left[ux'-vy'\right]dt+i\int_a^b\left[vx'+uy'\right]dt \tag{12}$$

Before making use of the concept of the complex differential given by equation (11) into equation (12), one can re-write the last one as follows:

$$\int_C f(z)dz=\int_C udx-vdy+i\int_C vdx+udy \tag{13}$$

Example

Evaluate $\int_{C_i} zdz, i=1,2$

The first path C_1 is the line joining the two points $(-1-i)$ and $(3+i)$, while C_2 is the parabola $x=y^2+2y$ joining the two points $(-1-i)$ and $(3+i)$, as shown in Fig. **(14)**.

Let us first find the equation of the first path in Cartesian coordinates, leads to:

$$x=2y+1$$

Now let us choose the parametrization for the straight line, by assuming:

$$y=t$$

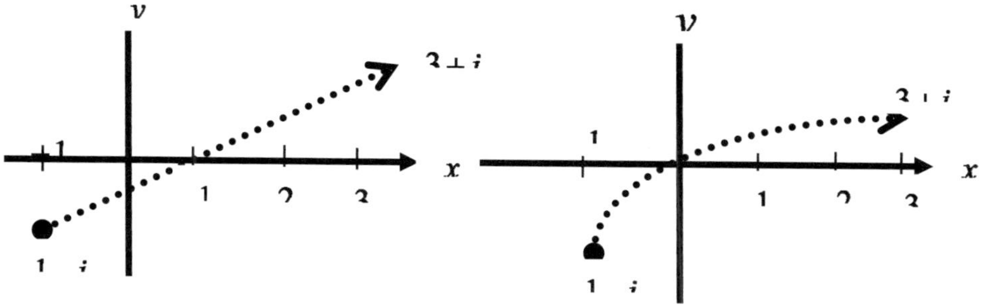

Figure 14: The two paths joining the two points.

Therefore the parameterization of the straight line becomes:

$$C_1 : z(t) = 2t + 1 + it$$
$$\Rightarrow$$
$$dz = (2 + i)\,dt$$

We have now $f\left(z(t)\right) = 2t + 1 + it$, if we apply the above theorem to the integral over C_1, then:

$$\int_{C_1} z\,dz = \int_{C_1} (2t + 1 + it)(2 + i)\,dt$$

Now then:

$$\int_{C_1} z\,dz = \int_{C_1} (3t + 2)\,dt + i\int_{C_1} (4t + 1)\,dt = 4 + 2i$$

A similar procedure will be done for the parabola, then the equation of the parabola takes the form $x = y^2 + 2y$ and so:

$$C_2 : z(t) = t^2 + 2t + it \quad \& \quad dz = (2t + 2 + i)\,dt \text{ for } -1 \le t \le 1$$

Therefore

$$\int_{C_2} z\,dz = \int_{-1}^{1}\left(t^2 + 2t + it\right)(2t + 2 + i)\,dt = \int_{-1}^{1}\left(2t^3 + 6t^2 + 3t\right)dt + i\int_{-1}^{1}\left(3t^2 + 4t\right)dt$$
$$= 4 + 2i$$

It is clear that the two results are the same, but this outcome does not hold in general.

3.11. GOURSAT THEOREM OF INTEGRATIONS

The integration of a complex function over a certain domain will be zero if and only if the function is analytic **[24]**.

3.12. GREEN'S THEOREM

Green's theorem relates line integral for a complex function over a closed and simple closed path and the double integration over the region bounded by the path. The theorem states that if C is a simple and closed path around a region R, and if P and Q are two continuous functions and their partial derivatives are exist, then [25-26]:

$$\int_C P(x,y)\,dx + Q(x,y)\,dy = \iint_R Q_x(x,y) - P_y(x,y)\,dxdy$$

Theorem

Suppose that we have two simple closed curves and both are positively oriented and one of them inside the other, say C_1 *inside* C_2. Suppose that the complex function is analytic in a domain D contains the two paths, as shown in Fig. (**15**), then **[27]**:

$$\int_{C_1} f(z)\,dz = \int_{C_2} f(z)\,dz$$

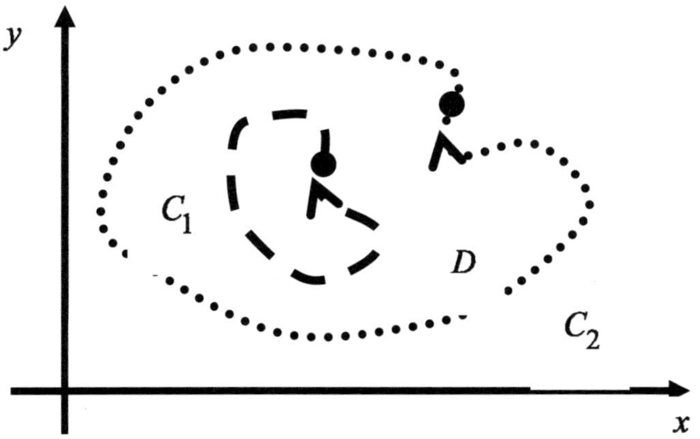

Figure 15: The domain D and the simple closed contours C_1 and C_2.

Corollary

Suppose that C is a simple closed contour with positive orientation and let z_0 be a fixed point lies interior to C, then:

$$\int_C \frac{dz}{z - z_0} = 2\pi i$$

and

$$\int_C \frac{dz}{\left(z - z_0\right)^n} = 0\,,\ n \text{ is an integer} \neq 1$$

3.13. GENERALIZED GOURSAT THEOREM

Suppose that a positive oriented curve has inside it a number n of closed and positive oriented curves, shown in Fig. (**16**), **[28]** then:

$$\int_C f(z)\,dz = \sum_{k=1}^{n} \int_{C_k} f(z)\,dz$$

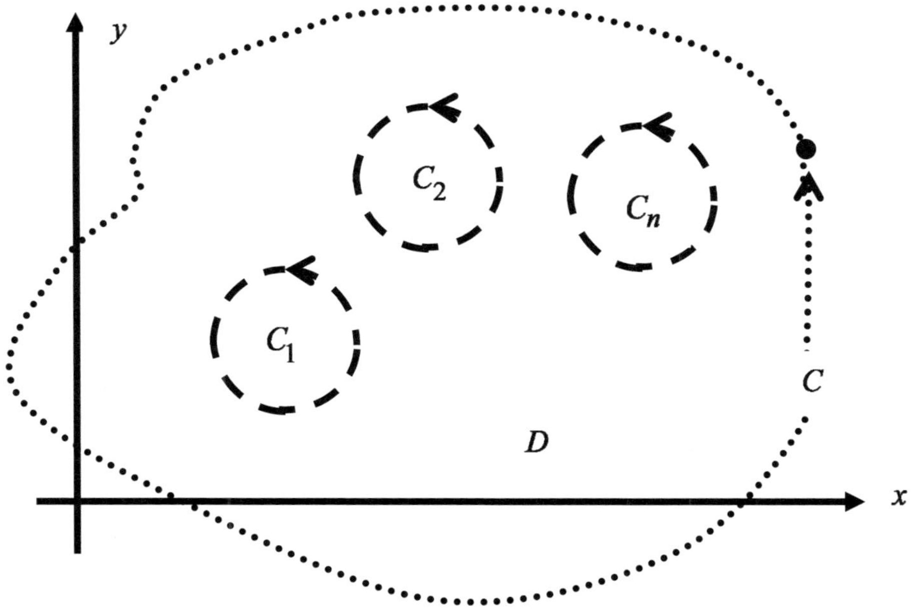

Figure 16: The multiply connected domain D and the contours $C, C_1, C_2, \ldots\ldots, C_n$.

Example

Show that $\int_{C_2^+(0)} \frac{2z}{z^2+2} dz = 4\pi i$

Solution

The path $C_2^+(0)$ is the circle $\{z : |z| = 2\}$ with positive orientation. Let us make use of partial fraction to the integrand, gives:

$$\frac{2z}{z^2+2} = \frac{2z}{\left(z+i\sqrt{2}\right)\left(z-i\sqrt{2}\right)} = \frac{1}{\left(z+i\sqrt{2}\right)} + \frac{1}{\left(z-i\sqrt{2}\right)}$$

Therefore

$$\int_{C_2^+(0)} \frac{2z}{z^2+2} dz = \int_{C_2^+(0)} \frac{1}{\left(z+i\sqrt{2}\right)} dz + \int_{C_2^+(0)} \frac{1}{\left(z-i\sqrt{2}\right)} dz \qquad \text{(A)}$$

The points $z = \pm i\sqrt{2}$ lie interior to $C_2^+(0)$, so by using corollary before, implies that

$$\int_{C_2^+(0)} \frac{1}{\left(z \pm i\sqrt{2}\right)} dz = 2\pi i$$

Substituting these values into equation (a), leads to:

$$\int_{C_2^+(0)} \frac{2z}{z^2+2} dz = 4\pi i$$

Example

Show that $\int_{C_1^+(i)} \frac{2z}{z^2+2} dz = 2\pi i$

Solution

The path $C_1^+(i)$ is the circle $\{z:|z-i|=1\}$ with positive orientation. Let us make use of partial fraction to the integrand, gives:

$$\frac{2z}{z^2+2}=\frac{2z}{\left(z+i\sqrt{2}\right)\left(z-i\sqrt{2}\right)}=\frac{1}{\left(z+i\sqrt{2}\right)}+\frac{1}{\left(z-i\sqrt{2}\right)}$$

$$\int_{C_2^+(0)}\frac{2z}{z^2+2}dz=\int_{C_2^+(0)}\frac{1}{\left(z+i\sqrt{2}\right)}dz+\int_{C_2^+(0)}\frac{1}{\left(z-i\sqrt{2}\right)}dz \qquad \text{(B)}$$

In this example, the point $z=+i\sqrt{2}$ lies interior to $C_1^+(i)$, but $z=-i\sqrt{2}$ does not using the above corollary, one can conclude that the second integral on the right side of equation (B) has the value $2\pi i$. The first integral in the same equation equals zero by the Goursat theorem because the function $f(z)=\dfrac{1}{\left(z+i\sqrt{2}\right)}$ is analytic on a simply connected domain $C_1^+(i)$.

Thus,

$$\int_{C_1^+(i)}\frac{2z}{z^2+2}dz=0+2\pi i=2\pi i$$

Exercises

1. Show that $\int_0^{\frac{\pi}{2}}\exp(t+it)\,dt=\frac{1}{2}\left(e^{\frac{\pi}{2}}-1\right)+\frac{i}{2}\left(e^{\frac{\pi}{2}}+1\right)$

2. Use the properties of complex integration to evaluate the following integrals:

(a) $\int_0^1(3t-i)^2\,dt$

(b) $\int_0^1(t+2i)^3\,dt$

(c) $\int_0^{\frac{\pi}{2}}\cosh(it)\,dt$

(d) $\int_0^2 \left(\frac{t}{t+i}\right) dt$

3. Show that $\int_{C_1} \bar{z} dz = -\pi i$ but $\int_{C_2} \bar{z} dz = -4i$

4. Give a parametrization of each of the following contour:

(a) $C = C_1 + C_2$ as shown in Fig. (**17**).

(b) $C = C_1 + C_2 + C_3$ as shown in Fig. (**18**).

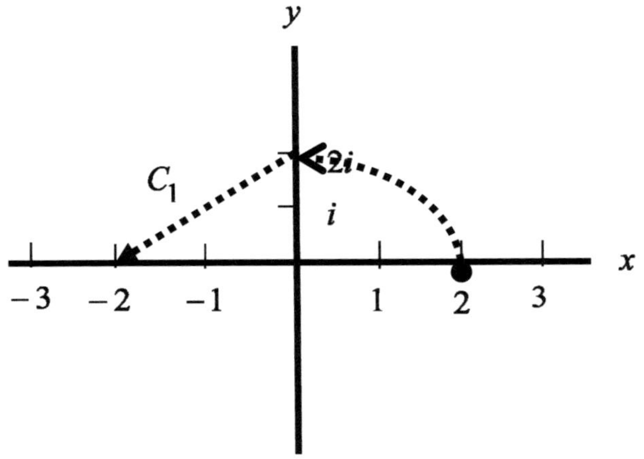

Figure 17: Problem (4-a).

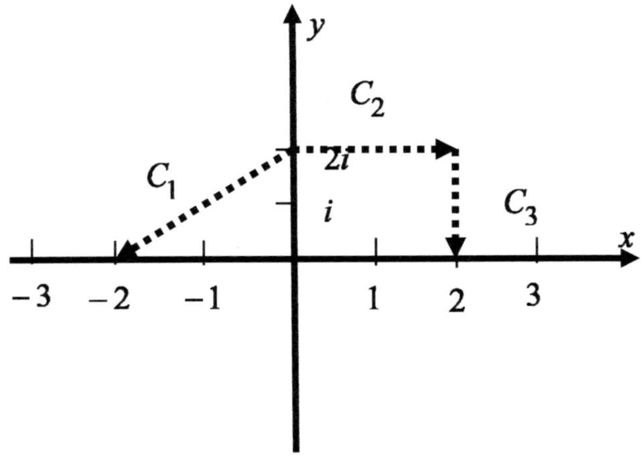

Figure 18: Problem (4-b).

5. Sketch the following curves:

(a) $z(t) = t^2 - 1 + i(t+4)$ for $1 \le t \le 3$

(b) $z(t) = \sin t + i\cos 2t$ for $-\dfrac{\pi}{2} \le t \le \dfrac{\pi}{2}$

(c) $z(t) = 5\cos t - i3\sin t$ for $\dfrac{\pi}{2} \le t \le 2\pi$

6. Consider the integral $\int_C z^2 dz$, where C is the positively oriented upper semi-circle of radius 1 and centered at the origin. Compute the integral exactly by selecting a parametrization for C.

7. Determine the domain of analyticity for the following functions and evaluate $\int_{C_1^+(0)} f(z)dz$

(a) $f(z) = \dfrac{z}{z^2+2}$

(b) $f(z) = \dfrac{1}{z^2+2z+2}$

8. Show that $\int_C \frac{1}{z} dz = 2\pi i$ in the square with vertices $1 \pm i$ and $-1 \pm i$ and having positive orientation.

9. Show that $\displaystyle\int_{C_1^+(0)} \frac{1}{4z^2-4z+5} dz = 0$

10. Find $\displaystyle\int_C \left(\frac{2z-1}{z^2-z}\right) dz$ for:

(a) Circle $C = C_2^+(0) = \{z : |z| = 2\}$ having positive orientation

(b) Circle $C = C_2^+(0) = \left\{z : |z| = \frac{1}{2}\right\}$ having positive orientation

11. Express e^z in the form $u + iv$ for the following values of z:

(a) $z = \frac{-\pi}{3}$

(b) $z = \frac{1}{2} - i\frac{\pi}{4}$

(c) $z = -4 + 5i$

(d) $z = -1 + i\frac{3\pi}{2}$

(e) $z = 1 + i\frac{5\pi}{4}$

(f) $z = \frac{\pi}{3} - 2i$

12. Find all values of z for which the following equations hold:

(a) $e^z = -4$

(b) $e^z = 2 + 2i$

(c) $e^z = \sqrt{3} - i$

(d) $e^z = 1 + i\sqrt{3}$

13. Express $\exp\left(z^2\right)$ and $\exp\left(\frac{1}{z}\right)$ in the Cartesian form $u(x, y) + iv(x, y)$

14. Find all values for:

(a) $\operatorname{Log}\left(ie^{2}\right)$

(b) $\operatorname{Log}\left(\sqrt{3}-i\right)$

(c) $\operatorname{Log}\left(i\sqrt{2}-\sqrt{2}\right)$

(d) $\operatorname{Log}\left[(1+i)^{4}\right]$

(e) $\log(-3)$

(f) $\log(8)$

(g) $\log(4i)$

(h) $\log\left(-\sqrt{3}-i\right)$

15. Find all values of z for which each equation holds:

(a) $\operatorname{Log}(z)=1-i\dfrac{\pi}{4}$

(b) $\operatorname{Log}(z-1)=i\dfrac{\pi}{2}$

(c) $\exp(z)=-ie$

(d) $\exp(z+1)=i$

Chapter 4: Complex Series

CHAPTER 4

Complex Series

Abstract: Mathematicians considered the contents of the present chapter continuity of the previous chapters. The last chapter was taking about theories of evaluating complex integrals. Herein, we will continue this subject, but firstly we focus the lights on some definitions and theories that are usually useful in studying theory of residues.

Keywords: Complex series, Laurent complex series, Maclaurin complex series, Taylor complex series.

4.1. THE COMPLEX FUNCTION

The disk of radius r and of center at the point $z=\alpha$ is defined as $D_r(\alpha)=\{z:|z-\alpha|\le r\}$, if there is an analytical function inside the disk, then the function will has a Taylor expansion and will converge to α **[29]**.

4.2. TAYLOR SERIES OF THE COMPLEX FUNCTION

If the complex function $f(z)$ is an analytic function at $z=\alpha$, then its Taylor series will take the following form:

$$f(\alpha)+(z-\alpha)f'(\alpha)+\frac{(z-\alpha)^2}{2!}f^{(2)}(\alpha)+\frac{(z-\alpha)^3}{3!}f^{(3)}(\alpha)..$$
$$=\sum_{k=0}^{\infty}\frac{(z-\alpha)^k}{k!}f^{(k)}(\alpha) \tag{1}$$

If the point of expansion is zero, then Taylor series takes the name Maclaurin series.

Lemma

If z, z_0, z_α are complex numbers with $z \neq z_0$ and $z \neq z_\alpha$, then:

$$\frac{1}{z-z_0}=\frac{1}{z-\alpha}+\frac{(z_0-\alpha)}{(z-\alpha)^2}+\frac{(z_0-\alpha)^2}{(z-\alpha)^3}+\frac{(z_0-\alpha)^3}{(z-\alpha)^4}$$
$$+\ldots+\frac{(z_0-\alpha)^n}{(z-\alpha)^{n+1}}+\left(\frac{1}{z-z_0}\right)\frac{(z_0-\alpha)^{n+1}}{(z-\alpha)^{n+1}} \tag{2}$$

S. G. Ahmed

Theorem: Taylor's theorem

Suppose that f is an analytic in a domain G and that $D_R(\alpha)=\{z:|z-\alpha|\le R\}$ is any disk in G. Then the Taylor series for f converges to $f(z)$ for all z in $D_R(\alpha)$; that is **[30]**:

$$f(z)=\sum_{k=0}^{\infty}\frac{(z-\alpha)^k}{k!}f^{(k)}(\alpha), \quad \text{for all } z\in D_R(\alpha) \qquad (3)$$

Example

Show that, for

$z\in D_1(0)$

$$(i)\quad \frac{1}{1-z^2}=\sum_{n=0}^{\infty}z^{2n} \qquad (4)$$

$$(ii)\quad \frac{1}{1+z^2}=\sum_{n=0}^{\infty}(-1)^n z^{2n}$$

Solution

For $z\in D_1(0)$

$$\frac{1}{1-z}=\sum_{n=0}^{\infty}z^n \qquad (5)$$

If we consider z^2 as z in equation (4), we will get the following equation:

$$\frac{1}{1-z^2}=\sum_{n=0}^{\infty}\left(z^2\right)^n=\sum_{n=0}^{\infty}z^{2n} \text{ for } z\in D_1(0)$$

But

$$z^2\in D_1(0) \text{ if and only if } z\in D_1(0).$$

Considering z^2 as z in equation (5) gives the second part of the required solution.

Example

For $f(z) = \sin^3 z$, find its Maclaurin expansion

Solution

As a usual manner, when it is required to find Taylor or Maclaurin expansions, it is required to find sufficient number of derivatives for the given function. Because both Taylor and Maclaurin for simple trigonometric functions, and because the given trigonometric function is of power more than one, this will lead us to make use the following trigonometric identity:

$$f(z) = \sin^3 z = \frac{3}{4}\sin z - \frac{1}{4}\sin 3z$$

But Maclaurin expansion for $f(z) = \sin z$ is well known as:

$$\sin z = \sum_{n=0}^{\infty}(-1)^n \frac{z^{2n+1}}{(2n+1)!}$$

Substituting into the identity given before, yields:

$$\sin^3 z = \frac{3}{4}\sum_{n=0}^{\infty}(-1)^n \frac{z^{2n+1}}{(2n+1)!} - \frac{1}{4}\sum_{n=0}^{\infty}(-1)^n \frac{(3z)^{2n+1}}{(2n+1)!}$$

$$\sin^3 z = \frac{3}{4}\sum_{n=0}^{\infty}(-1)^n \frac{z^{2n+1}}{(2n+1)!} - \frac{3}{4}\sum_{n=0}^{\infty}(-1)^n \frac{9^n (z)^{2n+1}}{(2n+1)!}$$

$$\sin^3 z = \sum_{n=0}^{\infty}(-1)^n \frac{3(1-9^n)z^{2n+1}}{4(2n+1)!}$$

The last expression is the Maclaurin series for the given trigonometric function $f(z)$.

4.3. EXPANSION OF FUNCTIONS WITH SINGULARITIES

In some cases, we encounter function has a singularity at certain point, but outside has not, *e.g.*, the function $f(z) = \frac{1}{z^3}e^z$ is not analytic when $z = 0$ but is analytic for $|z| > 0$. The function in its present form does not have a Maclaurin series. If we use the Maclaurin series for $g(z) = e^z$, however, and formally divide each term in that series by z^3, we obtain the representation:

$$f(z) = \frac{1}{z^3}e^z = \frac{1}{z^3} + \frac{1}{z^2} + \frac{1}{2!z} + \frac{1}{3!} + \frac{z}{4!} + \frac{z^2}{5!} + \frac{z^3}{6!} + \ldots\ldots$$

This is the expansion of the given function.

4.3.1. Annulus

The annulus is the region between two circles with the same center, mathematically defined as **[31]**:

$$A(\alpha, r, R) = \{z : r|z - \alpha| < R\}$$

One can define the annulus as: Given $0 \le r < R$, the annulus centered at α with radii r, R then the annulus is represented by the previous expression. From the above definition one can define the annulus mathematically as follows, see Fig. (**1**):

$$\overline{A}(\alpha, r, R) = \{z : r|z - \alpha| < R\}$$

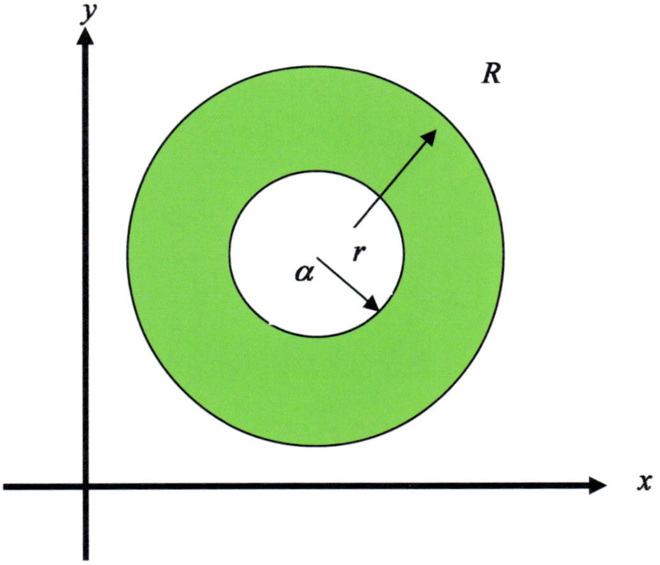

Figure 1: The shaded portion is the open annulus $A(\alpha, r, R)$.

4.3.2. Laurent Series for a Complex Function

The series $\sum_{n=-\infty}^{n=+\infty} c_n(z-\alpha)^n$, called a Laurent series, and can be defined as follows:

$$\sum_{n=-\infty}^{n=+\infty} c_n(z-\alpha)^n = \sum_{n=1}^{n=+\infty} c_{-n}(z-\alpha)^{-n} + c_0 + \sum_{n=1}^{n=+\infty} c_{+n}(z-\alpha)^{+n} \quad (6)$$

with complex numbers $c_n, 0, \pm 1, \pm 2, \ldots.$.

The expansion can take another simplified form as follows:

$$\sum_{n=0}^{n=+\infty} c_{+n}(z-\alpha)^{+n} = c_0 + \sum_{n=1}^{n=+\infty} c_{+n}(z-\alpha)^{+n}$$

Another form for the same expansion is as follows:

$$\sum_{n=-\infty}^{n=+\infty} c_{+n}(z-\alpha)^{+n} = \sum_{n=-\infty}^{n=-1} c_{+n}(z-\alpha)^{+n} + \sum_{n=0}^{n=-1} c_{+n}(z-\alpha)^{+n} \quad (7)$$

Theorem

The theorem simply states that if Laurent series converges on an annulus $A(\alpha, r, R)$, then the series converges on any sub-annulus $\overline{A}(\alpha, s, t)$, where $r < s < t < R$.

The next theorem related to theory of convergence and the corollary will be helpful for solving the next and any related problems.

Theorem

If $|z| < 1$, the series $\sum_{n=0}^{\infty} z^n$ converges to $f(z) = \frac{1}{1-z}$, that is, if $|z| < 1$, then:

$$\sum_{n=0}^{\infty} z^n = 1 + z + z^2 + \ldots + z^k + \ldots. = \frac{1}{1-z} \quad (8)$$

and if $|z| \geq 1$, the series diverges

Corollary

If $|z|>1$, the series $\sum_{n=0}^{\infty} z^{-n}$ converges to $f(z)=\frac{1}{z-1}$, that is, if $|z|>1$, then:

$$\sum_{n=0}^{\infty} z^{-n} = z^{-1}+z^{-2}+\ldots+z^{-n}+\ldots.=\frac{1}{z-1} \tag{9}$$

or equivalently,

$$-\sum_{n=0}^{\infty} z^{-n} = -z-z^{-1}-z^{-2}\ldots-z^{-n}-\ldots.=\frac{1}{1-z} \tag{10}$$

and if $|z|\le 1$, the series diverges

Example

For $f(z)=\frac{3}{2+z-z^2}$. Find all possible Laurent series.

Solution

The first step is to determine the points of singularities. The function $f(z)$ has singularities at $z=-1,2$. Therefore, the given function is analytic in the following:

1. The disk $D:|z|<1$
2. The annulus $A:1|z|<2$
3. The region $R:|z|>2$

To find all possible Laurent series, let us start dealing with the given complex function by making use of partial fractions:

$$f(z)=\frac{3}{2+z-z^2}=\frac{3}{(1+z)(2-z)}=\frac{1}{(1+z)}+\frac{1}{2}\frac{1}{\left(1-\frac{z}{2}\right)} \tag{11}$$

Make use of equations (8-10) to obtain the representations in terms on the right-hand side of equation (11), leads to:

$$\frac{1}{1+z}=\sum_{n=0}^{\infty}(-1)^n z^n \qquad \text{valid for } |z|<1 \tag{12}$$

$$\frac{1}{1+z}=\sum_{n=0}^{\infty}\frac{(-1)^{n+1}}{z^n} \qquad \text{valid for } |z|>1 \tag{13}$$

$$\frac{1}{2}\left(\frac{1}{1-\frac{z}{2}}\right)=\sum_{n=0}^{\infty}\frac{z^n}{2^{n+1}} \qquad \text{valid for } |z|<2 \tag{14}$$

$$\frac{1}{2}\left(\frac{1}{1-\frac{z}{2}}\right)=\sum_{n=1}^{\infty}\frac{-2^{n-1}}{z^n} \qquad \text{valid for } |z|>2 \tag{15}$$

Note that the representations (12) and (14) are valid in the disk D, therefore:

$$f(z)=\sum_{n=1}^{\infty}\left[(-1)^n+\frac{1}{2^{n+1}}\right]z^n \qquad \textbf{valid for } |z|<1$$

is the Laurent series. After finding the Laurent series in the disk, next is to find the expansion in the annulus. The expansions given by equations (13) and (15) are valid in the annulus and can be re-written as follows:

$$f(z)=\sum_{n=1}^{\infty}\frac{(-1)^{n+1}}{2^n}+\sum_{n=0}^{\infty}\frac{z^n}{2^{n+1}} \qquad \text{valid for } 1<|z|<2 \tag{16}$$

Finally; one can use expansions given by equations (13) and (15), to find the expansion that will be valid in the region, as follows:

$$f(z)=\sum_{n=1}^{\infty}\frac{(-1)^{n+1}-2^{n-1}}{z^n} \qquad \text{valid for } |z|>2 \tag{17}$$

Example

Find Laurent series for $f(z) = \dfrac{\cos z - 1}{z^4}$

Solution

The first step is to Maclaurin series for $\cos z - 1$, as follows:

$$\cos z - 1 = \frac{-1}{2!}z^2 + \frac{1}{4!}z^4 - \frac{1}{6!}z^6 + \ldots.$$

Then:

$$f(z) = \frac{\cos z - 1}{z^4} = \frac{\frac{-1}{2!}z^2 + \frac{1}{4!}z^4 - \frac{1}{6!}z^6 + \ldots.}{z^4}$$

Divide each term by z^4, leads to:

$$f(z) = \frac{-1}{2z^2} + \frac{1}{24} - \frac{z^2}{720} + \ldots \text{ valid for } z \neq 0$$

4.4. SINGULARITY

The point $z = \alpha$ is called point of singularity, or singularity, of the complex function $f(z)$ if $f(z)$ is not analytic at that point **[32]**.

4.4.1. Isolated Singularity

The isolated singularity of a complex function is defined as the point $z = \alpha$ at which, the function is not analytic but there exists positive real number R such that the function is analytic everywhere in the disk $D_R^*(\alpha)$. An illustrated example, $f(z) = \dfrac{1}{1-z}$ has an isolated singularity at $z = \alpha = 1$. It is important now to say that the function of isolated singularity always have Laurent expansion.

4.4.2. Classification of Singularities

As mentioned earlier any complex function with isolated singularity will have Laurent series, therefore, let the complex function $f(z)$ has an isolated singularity at the point α, then its Laurent series will be:

$$f(z)=\sum_{n=-\infty}^{n=+\infty} c_{+n}(z-\alpha)^{+n} \text{ valid for all } z\in A(\alpha,0,R) \tag{18}$$

Therefore, two different singularities can be distinguished as follow:

4.4.2.1. Removable Singularity

1. If $c_{+n}=0$, for $n=-1,-2,-3,...$, then $f(z)$ has a removable singularity at α

4.4. 2.2. Pole of Order k

2. If k, is a positive integer such that $c_{-k}\neq 0$, but $c_n=0$, **for** $n<-k$, then the complex function has a ***pole of order*** k at the point α.

4.4.2.3. Essential Singularity

3. If $c_{+n}\neq 0$,, for infinitely many negative integers n, then $f(z)$ has an ***essential singularity*** at α.

Example

Consider the complex function $f(z)=\dfrac{\sin z}{z}$ which has an isolated singularity at $z=0$, therefore the Laurent series for $f(z)$ will be:

$$\begin{aligned} f(z)&=\frac{\sin z}{z}=\frac{1}{z}\left(z-\frac{z^3}{3!}+\frac{z^5}{5!}-+.......\right)\\ &=1-\frac{z^2}{3!}+\frac{z^4}{5!}-\frac{z^6}{7!}+-....\text{valid for} |z|>0 \end{aligned} \tag{19}$$

If we define $f(0)=1$, for then $f(z)$ will be analytic at $z=0$.

Example

Consider the complex function $f(z)=\dfrac{\cos z-1}{z^2}$ which has an isolated singularity at $z=0$, therefore the Laurent series for $f(z)$ will be:

$$f(z)=\frac{\cos z-1}{z^2}=\frac{1}{z^2}\left(-\frac{z^2}{2!}+\frac{z^4}{4!}-\frac{z^6}{6!}+.......\right)$$
$$=-\frac{1}{2}+\frac{z^2}{4!}-\frac{z^4}{6!}+-....\text{valid for}\,|z|>0 \tag{20}$$

If we define $f(0)=-\frac{1}{2}$, then $f(z)$ will be analytic for all z

Example

The function $f(z)=\frac{\sin z}{z^3}=\frac{1}{z^2}-\frac{1}{3!}+\frac{z^2}{5!}-\frac{z^4}{7!}+.......$ has a pole of order 2 at the point $z=0$

Example

The function $f(z)=\frac{e^z}{z}=\frac{1}{z}+1+\frac{z}{2!}+\frac{z^2}{3!}+.......$ has a simple pole at the point $z=0$

Example

The function $f(z)=z^2\sin\frac{1}{z}=z-\frac{1}{3!}z^{-1}+\frac{z^{-3}}{5!}-\frac{z^{-5}}{7!}+.......$ has an essential singularity at the origin.

Theorem

A function $f(z)$ analytic in $D_R(\alpha)$ has a zero of order k at the point α if and only if its Taylor series given by

$$f(z)=\sum_{n=0}^{n=+\infty}c_{+n}(z-\alpha)^{+n} \tag{21}$$

$$c_0=c_1=c_2=...........=c_{k-1}=0 \text{ but } c_k\neq 0 \tag{22}$$

Example

Let us investigate the concept of zero by analyzing the function $f(z)=z\sin z^2$ has a zero of order 3 at the point $z=0$. Definition of zero of order k confirms this fact because:

$$f^{/}(z)=2z^{2}\cos z^{2}+\sin z^{2}$$
$$f^{//}(z)=6z\cos z^{2}-4z^{3}\sin z^{2}$$
$$f^{///}(z)=6\cos z^{2}-8z^{4}\sin z^{2}-24z^{2}\sin z^{2}$$

Then

$$f(0)=f^{/}(0)=f^{//}(0)=0$$

but

$$f^{///}(0)=6\neq 0$$

This example leads us to the following theorem.

Theorem

The complex function will has zero of order k if and only if the following expression satisfied:

$$f(z)=(z-\alpha)^{k}g(z) \tag{23}$$

where $g(z)$ is analytic at the point α and $g(\alpha)\neq 0$

Exercises

1. Show that $\frac{1}{(1-z)^{2}}=\sum_{n=0}^{\infty}(n+1)z^{n}$ is valid for $z\in D_{1}(0)$

2. Find the Maclaurin series for:

(a) $f(z)=\cos^{3}z$. Use the trigonometric identity

$\cos^{3}z=\frac{1}{4}(\cos 3z+3\cos z)$

(b) $f(z)=(z^{2}+1)\sin z$

(c) $f(z)=e^{z}\cos z$. Note: $\cos z=\frac{1}{2}\left(e^{iz}+e^{-iz}\right)$ so $f(z)=\frac{1}{2}e^{(1+i)z}+\frac{1}{2}e^{(1-i)z}$.

Now use the Maclaurin series for e^{z}.

3. Find the Taylor series centered at $a = 1$ for:

(a) $f(z) = \dfrac{1-z}{z-2}$

(b) $f(z) = \dfrac{1-z}{z-3}$

(c) Hint: $\dfrac{1-z}{z-3} = \left(\dfrac{1}{2}\right)\dfrac{(z-1)}{\left(1-\left(\dfrac{z-1}{2}\right)\right)} = \left(\dfrac{1}{2}\right)(z-1)\dfrac{1}{\left(1-\left(\dfrac{z-1}{2}\right)\right)}$

4. Show that $f(z) = \dfrac{1}{(1-z)}$ has its Taylor series representation about the point $a = i$ given by:

(a) $f(z) = \sum_{n=0}^{\infty} \dfrac{(z-i)^n}{(1-i)^{n+1}}$ for all $z \in D_{\sqrt{2}}(i) = \{z : |z-i| < \sqrt{2}\}$

5. Find two Laurent series expansion for $f(z) = \dfrac{1}{z^3 - z^4}$ that involve powers of z.

6. Find the Laurent series for $f(z) = \dfrac{\sin 2z}{z^4}$ that involve powers of z.

7. Show that $\dfrac{1-z}{z-2} = -\sum_{n=0}^{\infty} \dfrac{1}{(z-1)^n}$ is valid for $|z-1| > 2$.

8. Locate the poles of $g(z) = \dfrac{1}{5z^4 + 26z^2 + 5}$ and specify their order.

9. Locate the poles of $g(z) = \dfrac{\pi \cot(\pi z)}{z^2}$ and specify their order.

10. Locate the zeros of the following functions:

(a) $f(z) = (1 + z^2)^4$

(b) $f(z) = \sin^2 z$

(c) $f(z) = z^2 + 2z + 2$

(d) $f(z) = z^3 \exp(z - 1)$

(e) $f(z) = z^3 \cos^2 z$

11. Locate the poles of the following functions and determine their order:

(a) $f(z) = (z^2 + 1)^{-3} (z - 1)^{-4}$

(b) $f(z) = z^{-1} (z^2 - 2z + 2)^{-2}$

(c) $f(z) = (z^6 + 1)^{-1}$

(d) $f(z) = z \cot z$

(e) $f(z) = z^{-5} \sin z$

12. Locate the singularities of the following functions and determine their type:

(a) $f(z) = \dfrac{z^2}{z - \sin z}$

(b) $f(z) = \sin\left(\dfrac{1}{z}\right)$

(c) $f(z) = z \exp\left(\dfrac{1}{z}\right)$

(d) $f(z)=\left(z+z^2\right)^{-1}\sin z$

(e) $f(z)=\dfrac{\cos z-\cos 2\pi}{z^4}$

CHAPTER 5

Theory of Residue

Abstract: You have now the necessary machinery to see some amazing applications of the tools that we had developed in the last few chapters. You will learn how Laurent expansions can give useful information concerning seemingly unrelated properties of complex functions. You will also learn how the ideas of complex analysis make the solution of very complicated integrals of real-valued functions as easy – valued functions as easy – literally – as the computation of complex integrals. The topic of our discussion in this chapter is one of the most important and most often used tools that applied scientists extract from the theory of complex variables. Residue theory is employed in a wide variety of applications ranging from the evaluation of real integrals to the stability of linear systems to image evaluation in photographic science. In this chapter, we develop and illustrate some of the fundamental techniques employed in complex integration by the use of residue theory.

Keywords: Application of residue theory to real definite integrals, improper integrals, the residue theorem.

5.1. THE RESIDUE THEOREM

Let $f(z)$ be an analytic function and suppose that z_0 is an isolated singularity of $f(z)$. Then, by definition, $f(z)$ is analytic in a deleted neighborhood of z_0. Then, $f(z)$ has a Laurent expansion that converges to $f(z)$ for all z in the dashed region shown in (Fig. **1**) **[33]**.

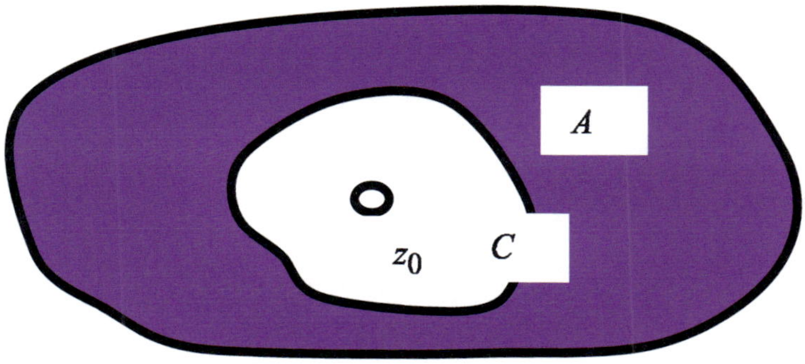

Figure 1: Configuration of the residue at the point of singularity.

One can write Laurent series in the following form:

$$f(z) = \sum_{n=-\infty}^{n=+\infty} c_n (z - z_0)^n$$

S. G. Ahmed

Definition

Let $f(z)$ be a function having a non-removable isolated singularity at the point $z = z_0$. Then $f(z)$ has the Laurent series representation for all z in some disk $D_R^*(z_0)$ given by $f(z) = \sum_{n=-\infty}^{n=+\infty} a_n (z - z_0)^n$. The coefficient a_{-1} is called the residue of $f(z)$ at the point $z = z_0$. In such a case the coefficients are given by:

$$c_n = \frac{1}{2\pi i} \int_C \frac{f(z)}{(z - z_0)^{n+1}} dz, \; n = 0, \pm 1, \pm 2, \pm 3, \ldots\ldots\ldots$$

where c is any positively oriented simple closed path lying entirely in the dashed region. In a particular case, for $n = -1$, we have:

$$c_{-1} = \frac{1}{2\pi i} \int_C f(z) dz$$

The number c_{-1} is the coefficient of the term $\frac{1}{z - z_0}$ in the Laurent expansion and is called the residue of $f(z)$ at the point z_0, and will be denoted as $\mathrm{Res}[f(z), z_0]$. Mathematically, this can stated as:

$$c_{-1} = \mathrm{Res}[f(z), z_0]$$

Therefore, one can write the expression for the following integral as:

$$\int_C f(z) dz = (2\pi i) \mathrm{Res}[f(z), z_0]$$

5.2. RESIDUE THEOREM

Suppose that $f(z)$ is analytic on and interior a positively oriented simple closed path C, except at a finite number of points $z_1, z_2, z_3, \ldots.., z_n$ each of which is an isolated singularity of $f(z)$, then **[34]**:

$$\int_C f(z)dz = (2\pi i)\sum_{j=1}^{n} \mathrm{Res}[f(z), z_0]$$

Theorem

Suppose that $f(z)$ has a pole of order n at the point z_0, then:

$$\mathrm{Res}[f(z), z_0] = \frac{1}{(n-1)!}\lim_{z \to z_0} \frac{d^{(n-1)}}{dz^{(n-1)}}\{(z-z_0)^n f(z)\}$$

Example

Evaluate $\int_C \frac{e^z}{z^2} dz$, $C:|z|=1$, the path is a positively oriented circle with center at the origin and of radius of unit length.

Solution

The integrand has a pole at the point $z_0 = 0$ and of order 2, then:

$$\mathrm{Res}[f(z), z_0] = \frac{1}{(n-1)!}\lim_{z \to z_0} \frac{d^{(n-1)}}{dz^{(n-1)}}\{(z-z_0)^n f(z)\}, \quad n=2$$

$$\mathrm{Res}[f(z), z_0] = \frac{1}{(2-1)!}\lim_{z \to z_0} \frac{d}{dz}\left\{(z-z_0)^2 \frac{e^z}{z^2}\right\}$$

$$\mathrm{Res}[f(z), z_0] = \lim_{z \to z_0} \frac{d}{dz}\left\{z^2 \frac{e^z}{z^2}\right\} = \lim_{z \to z_0} \frac{d}{dz}\{e^z\} = \{e^z\}_{z=z_0=0} = 1$$

Then

$$\int_C \frac{e^z}{z^2} dz = (2\pi i)\mathrm{Res}[f(z), z_0] = (2\pi i)$$

Example

Use residue theorem to evaluate the integral

$\int_C f(z)dz,\ f(z)=\dfrac{1}{(z-1)(z+1)}$ and $C:|z|=3$ the path is a positively oriented circle with center at the origin and of radius of 3 units.

Solution

The integrand has 2 poles, they are $z_1=+1, z_2=-1$ as shown in Fig. (**2**), and this requires the evaluation of the residue at each pole, separately then applies the theory of residue.

$$\mathrm{Res}\left[f(z),z_1\right]=\frac{1}{(1-1)!}\lim_{z\to+1}\{(z-1)f(z)\}=\lim_{z\to+1}\left\{\frac{1}{(z+1)}\right\}=\frac{1}{2}$$

$$\mathrm{Res}\left[f(z),z_2\right]=\frac{1}{(1-1)!}\lim_{z\to-1}\{(z+1)f(z)\}=\lim_{z\to-1}\left\{\frac{1}{(z-1)}\right\}=-\frac{1}{2}$$

Then

$$\int_C f(z)dz=(2\pi i)\sum_{j=1}^{2}\mathrm{Res}\left[f(z),z_j\right]=(2\pi i)\left\{\frac{1}{2}-\frac{1}{2}\right\}=0$$

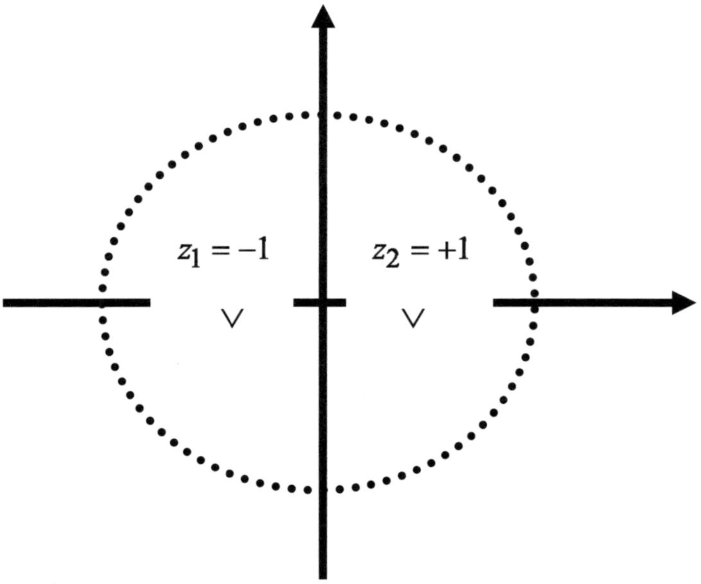

Figure 2: Configuration of the problem.

The following questions without answers and the reader should clearly know very their answers:

1. What is the difference between pole and zero?

2. If a function $f(z)$ has pole or zero, is it possible to say in both cases the function is analytic?

3. If the answer of 2nd question is yes, please say why?

Corollary

Suppose that:

1. $f(z)$ and $g(z)$ are both analytic at a point z_0

2. $f(z_0) \neq 0$

3. $g(z)$ has a zero of order 1 at the point z_0

Then

$$\mathrm{Res}\left[\frac{f(z)}{g(z)}, z_0\right] = \frac{f(z_0)}{g'(z_0)}$$

Example

Evaluate the residue of the rational function $h(z) = \dfrac{z^2 + 2z + 5}{z - i}$ at the point $z = i$

Solution

Both the numerator and the denominator of the given function is analytic at the point $z = i$, and the denominator has a zero of order 1 at that point. Also, the numerator does not vanish at that point, therefore:

$$\mathrm{Res}\left[h(z), z_0 = i\right] = \frac{f(z_0)}{g'(z_0)} = \left(\frac{z^2 + 2z + 5}{1}\right)_{z_0 = i} = 4 + 2i$$

5.3. APPLICATION OF RESIDUE THEORY TO REAL DEFINITE INTEGRALS

It is often necessary, especially in the various areas of applied mathematics, to employ methods of complex integration in order to evaluate certain types of real integrals that do not yield to the calculus of real variables. In the present section, we will study some of these integrals.

5.4. INTEGRALS OF RATIONAL FUNCTIONS OF SINE AND COSINE

Integrals of this type have the general form **[35]**:

$$I = \int_0^{2\pi} f(\cos t, \sin t)\, dt$$

where $f(\cos t, \sin t)$ is a rational function of $\cos t$ & $\sin t$, *i.e.*, it involves only integer powers of trigonometric functions and is finite $0 \le t \le 2\pi$. The basic substitution that is employed in the evaluation of such integrals is:

$$z = \cos t + i \sin t, \qquad 0 \le t \le 2\pi \tag{1}$$

Note: as t varies in the interval $0 \le t \le 2\pi$, z ***describes*** a unit circle $C: |z| = 1$ in the positive sense.

From equation (1), one can write the following relations:

$$\frac{1}{z} = \cos t - i \sin t$$

$$\cos t = \frac{1}{2}\left(z + \frac{1}{z}\right) \tag{2}$$

$$\sin t = \frac{1}{2i}\left(z - \frac{1}{z}\right) \tag{3}$$

Also

$$dz = (-\sin t + i \cos t)\, dt \Rightarrow dz = (i\, z)\, dt$$

Therefore

$$dz = (-\sin t + i\cos t)\,dt \Rightarrow dz = (i\,z)\,dt$$

$$dt = \frac{1}{i\,z}dz = \frac{-i}{z}dz$$

Clearly

$$dt = \frac{-i}{z}dz \tag{4}$$

Example

Evaluate the integral $\int_0^{2\pi} \frac{1}{2+\cos t}dt$

Solution

Start the solution by substituting equations (2-4) into the given integral, leads to:

$$I = \int_0^{2\pi} \frac{1}{2+\cos t}dt = -i\int_C \left(\frac{dz}{z\left[2 + \underbrace{\frac{1}{2}\left(z + \frac{1}{z} \right)}_{=\cos t} \right]} \right)$$

= dt

$$I = -2i\int_C \frac{dz}{z^2+4z+1}$$

$$I = -2i \int_C \frac{dz}{\left(z+2+\sqrt{3}\right)\left(z+2-\sqrt{3}\right)}$$

The path is a positively oriented circle with center at the origin and of unit radius. There are two points of singularities, they are:

$$z_1 = \left(-2-\sqrt{3}\right)$$
$$z_2 = \left(-2+\sqrt{3}\right)$$

Note: z_1 lies exterior the domain, therefore

$$\mathrm{Res}\left[f(z), z_1\right] = 0$$

$$\mathrm{Res}\left[f(z), z_2\right] = \lim_{z \to -2+\sqrt{3}} \left(\frac{1}{z+2+\sqrt{3}}\right) = \left(\frac{1}{2\sqrt{3}}\right)$$

$$I = -2i(2\pi i)\sum_{j=1}^{2} \mathrm{Res}\left[f(z), z_j\right] = -2i(2\pi i)\left[0 + \left(\frac{1}{2\sqrt{3}}\right)\right] = \left(\frac{2\pi}{\sqrt{3}}\right)$$

Example

Evaluate the integral $\int_0^{2\pi} \frac{\sin^2 t}{5-4\cos t} dt$

Solution

Start the solution by substituting equations (2-4) into the given integral, simplify, yields:

$$\int_0^{2\pi} \frac{\sin^2 t}{5-4\cos t} dt = \left(-\frac{i}{4}\right) \int_C \frac{z^4 - 2z^2 + 1}{z^2(z-2)(2z-1)} dz$$

The path is a positively oriented circle with center at the origin and of unit radius. There are three poles of singularities, they are:

$$z_1 = 0$$

$$z_2 = 2$$

$$z_3 = \frac{1}{2}$$

Next step is to evaluate the residues corresponding to the above three poles:

Note that: the 2nd pole is exterior the path C, therefore the corresponding residue equals zero. Now then, we will turn to evaluate the residues corresponding to the 1st and 3rd poles as follows:

$$\mathrm{Res}\left[f(z), z_1\right] = \frac{5}{4}$$

$$\mathrm{Res}\left[f(z), z_3\right] = -\frac{3}{4}$$

Then, the value of the given integral will be:

$$I = -\frac{i}{4}(2\pi i)\sum_{j=1}^{3}\mathrm{Res}\left[f(z), z_j\right] = -\frac{i}{4}(2\pi i)\left(\frac{5}{4} + 0 - \frac{3}{4}\right) = \frac{\pi}{4}$$

Note:

The above method can be generalized to the case of any integral of the form:

$$\int_0^{2\pi} \frac{dt}{a + b\cos t}$$

Or

$$\int_0^{2\pi} \frac{dt}{a + b\sin t}$$

Such that $|a| > |b|$

5.5. IMPROPER INTEGRALS

Case (1)

In this section, we will deal with integral in the following form:

$$I = \int_{-\infty}^{+\infty} f(x)\,dx, \quad f(x) = \frac{P(x)}{Q(x)}, \quad \text{order of } P(x) \le \text{order of } Q(x) - 2$$

Such that, $\lim_{x \to \pm\infty} xf(x) = 0$

Consider a complex function $f(x)$, which has a finite number of poles $z_1, z_2, z_3, .., z_n$ in the upper-half plane, we will ignore any pole in the lower-half plane. We now consider the following complex integral:

$$I = \int_C f(z)\,dz$$

where the path $C = C_1 + C_2$ is shown in Fig. (**3**).

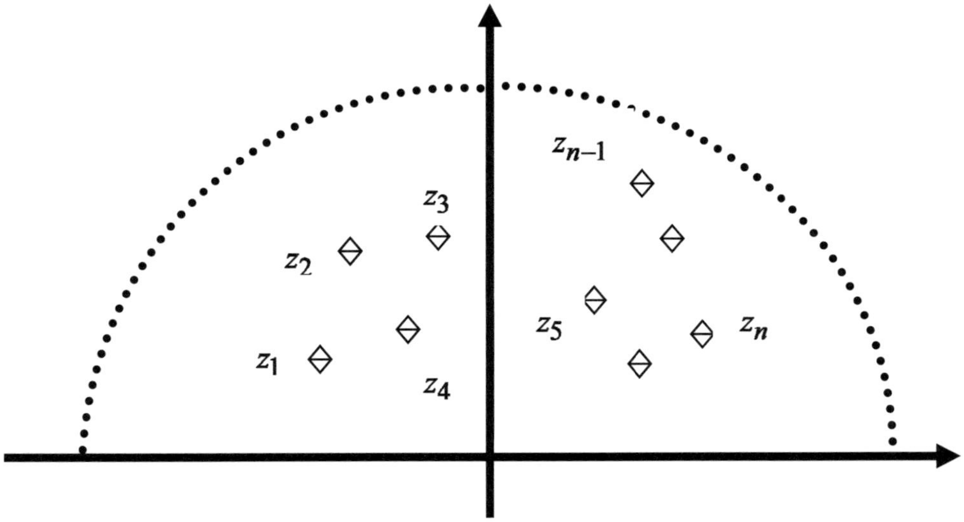

Figure 3: Different paths.

The path is as shown in Fig. (**3**) consists of the upper half of a semi-circle C_1 and the line segment C_2 coincide with the horizontal axis from $-R$ to $+R$. It is important to remember that all poles are interior of the region enclosed the path C.

In this contest, we will deal with:

$$I = \int_{-\infty}^{+\infty} f(x)\,dx = \oint_C f(z)\,dz = (2\pi i)\sum_{j=1}^{n} \mathrm{Re}\,s(f, z_j)$$

We have now four different cases of poles, they are shown clearly in Fig. **(4)**.

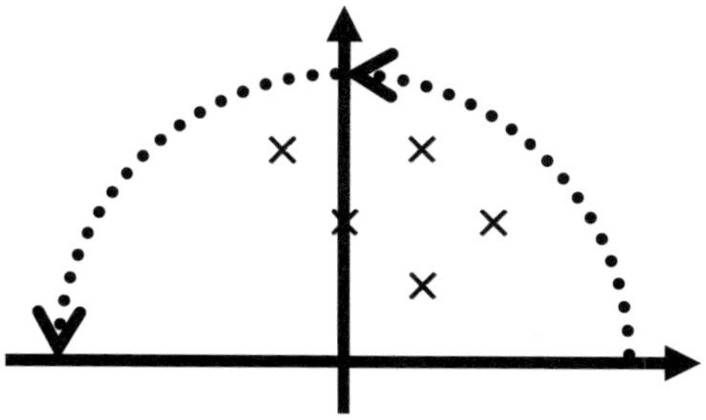

Figure 4a: Case (a) $f(z)$ has no poles on the real axis.

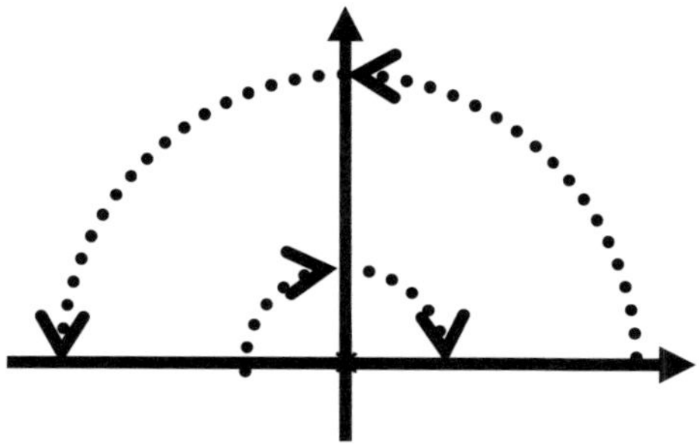

Figure 4b: Case (b) $f(z)$ has a real pole at $z = 0$.

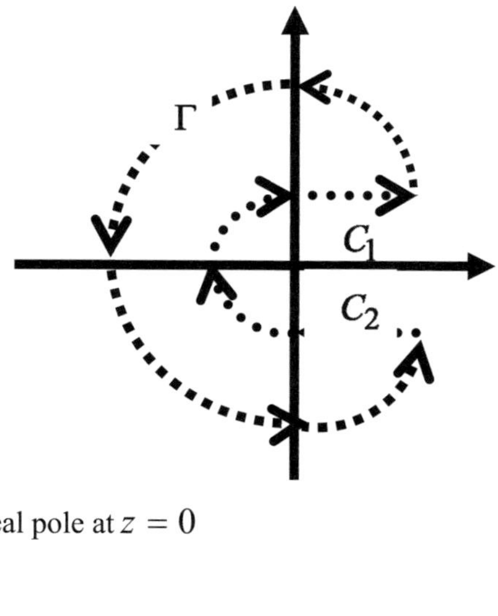

Figure 4c: Case (c)..has a real pole at $z = 0$

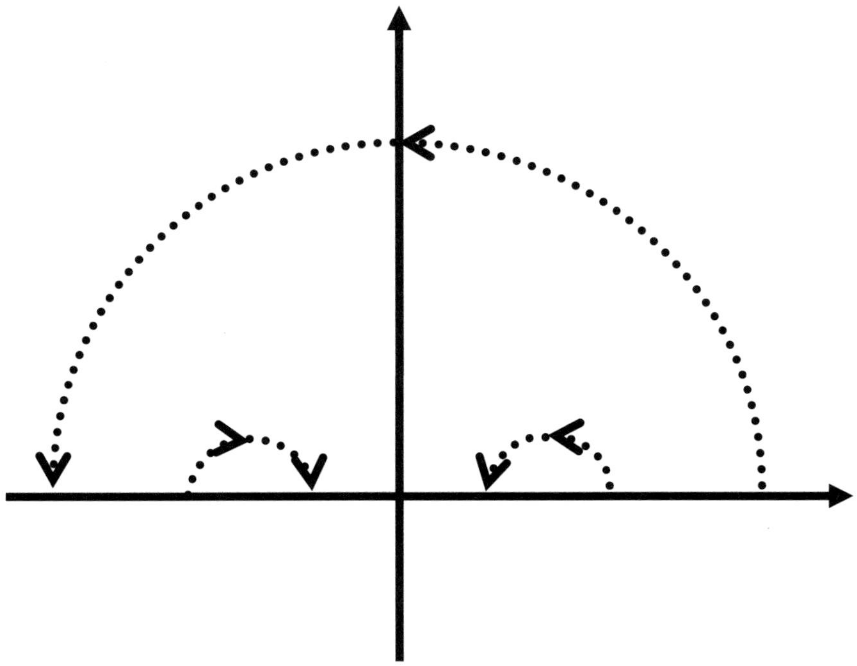

Figure 4d: Case (d) $f(z)$ has two real poles

In all the above four cases of poles existence, we want to evaluate an integral of the following form:

$$I = \int_{-\infty}^{+\infty} \frac{p(x)}{Q(x)} dx = \oint_C \frac{p(z)}{Q(z)} dz$$

In this case, Cauchy theorem will be applied as follows:

$$I = \int_{-\infty}^{+\infty} \frac{p(x)}{Q(x)} dx = \oint_C \frac{p(z)}{Q(z)} dz = (2\pi i) \sum_{j=1}^{n} \mathrm{Res}(f, z_j)$$

Example

Show that $I = \int_{-\infty}^{+\infty} \frac{dx}{\left(x^2+4\right)^2} = \frac{\pi}{16}$

Solution

First of all, let us remember something about the integrand:

$$f(x)=\frac{1}{\left(x^2+4\right)^2} \Leftrightarrow f(z)=\frac{1}{\left(z^2+4\right)^2}$$

The next step is to make some analysis, *i.e.*, to know if the function has poles, zeros or nothing. As we see, the function has two poles each of which is of order two, they are:

$$z_1=2i$$
$$z_2=-2i$$

We then take a path shown before in Fig. (**3**), with radius $R>2$, so that $z_1=2i$, which is the only singularity for the given function in the upper-half plane, and it will be interior of the path C. Now, let us start dealing with the given integral:

$$I=\int_C\frac{dz}{\left(z^2+4\right)^2}=\int_{C_1}\frac{dz}{\left(z^2+4\right)^2}+\int_{C_2}\frac{dz}{\left(z^2+4\right)^2}$$

The right-hand side of the above equation can take another form therefore; let us re-write the above equation in the following form:

$$I=\int_C\frac{dz}{\left(z^2+4\right)^2}=\int_{C_1}\frac{dz}{\left(z^2+4\right)^2}+\int_{-R}^{+R}\frac{dx}{\left(x^2+4\right)^2}$$

The second integral took the above form because the path coincides with the horizontal real x-axis. We now need to - evaluate the integral on the path c_1 - the residue is obtained so as we can apply onto the Cauchy theorem of integration.

$$\operatorname{Res}\left[f(z),2i\right]=\frac{1}{(2-1)!}\lim_{z\to 2i}\frac{d}{dz}\left[(z-2i)^2\frac{1}{\left(z^2+4\right)^2}\right]$$

$$\operatorname{Res}\left[f(z),2i\right]=\lim_{z\to 2i}\frac{d}{dz}\left[(z-2i)^2\frac{1}{(z-2i)^2(z+2i)^2}\right]$$

$$\text{Res}\left[f(z),2i\right]=\lim_{z\to 2i}\frac{d}{dz}\left[\frac{1}{(z+2i)^2}\right]$$

$$\text{Res}\left[f(z),2i\right]=\lim_{z\to 2i}\left[\frac{-2}{(z+2i)^3}\right]$$

$$\text{Res}\left[f(z),2i\right]=\left[\frac{1}{32i}\right]$$

Note the 2nd integral is equal to zero, therefore the value of the given original integral will be:

$$I=(2\pi i)\text{Res}\left[f(z),2i\right]=(2\pi i)\frac{1}{32i}=\frac{\pi}{16}$$

Example

Show that $I=\int_{-\infty}^{+\infty}\frac{x^2}{(x^2+1)^2(x^2+2x+2)}dx=\frac{7\pi}{50}$

Solution

First of all, let us consider something about the integrand:

$$f(x)=\frac{x^2}{(x^2+1)^2(x^2+2x+2)}\Leftrightarrow f(z)=\frac{z^2}{(z^2+1)^2(z^2+2z+2)}$$

The next step is to make some analysis, *i.e.*, to know if the function has poles, zeros or nothing. As we see, the function has two poles, one of which is of order two and the other is of order one, they are:

$$z_1=i, \quad \text{order 2}$$
$$z_2=-1-i \quad \text{order 1}$$

We then take a path shown before in Fig. (**3**), with radius $R>2$, so that both the two poles lies entirely interior C, Now, let us start dealing with the given integral:

$$I = \int_C \frac{z^2}{\left(z^2+1\right)^2\left(z^2+2z+2\right)} dz = \int_{C_1} \frac{z^2}{\left(z^2+1\right)^2\left(z^2+2z+2\right)} dz + \int_{C_2} \frac{z^2}{\left(z^2+1\right)^2\left(z^2+2z+2\right)} dz$$

We now need to - evaluate the integral on the path c_1 - the residue so as we can apply the Cauchy theorem of integration.

$$\text{Res}\left[f(z), i\right] = \lim_{z\to i}\left[(z-i)^2 \frac{z^2}{(z-i)^2(z+i)^2\left(z^2+2z+2\right)}\right] = \frac{9i-12}{100}$$

$$\text{Res}\left[f(z), -1+i\right] = \lim_{z\to -1+i}\left[(z+1-i)^2 \frac{z^2}{(z+1)^2(z+1-i)(z+1+i)}\right] = \frac{3-4i}{25}$$

Then

$$I = (2\pi i)\sum_{j=1}^{2} \text{Res}\left[f(z), z_j\right] = (2\pi i)\left[\frac{9i-12}{100} + \frac{3-4i}{25}\right] = \frac{7\pi}{50}$$

Improper Integrals

Case (2)

In this section, we will deal with integral in the following form:

$$I = \int_{-\infty}^{+\infty} g(x) e^{i\alpha x} dx, \quad f(x) = \frac{P(x)}{Q(x)}, \quad \text{order of } P(x) \le \text{order of } Q(x) - 2$$

Such that, $\alpha > 0$

Under these conditions, the corresponding complex function $f(z) = g(z)e^{i\alpha z}$, the function $f(x)$, has a finite number of poles $z_1, z_2, z_3, .., z_n$ in the upper-half plane we will ignore any pole in the lower-half plane. Then it can be shown that:

$$I = \int_{-\infty}^{+\infty} g(x) e^{i\alpha x} dx = \oint_C g(z) e^{i\alpha z} dz = (2\pi i)\sum_{j=1}^{n} \text{Res}\left(f, z_j\right)$$

The above result lends itself to the evaluation of certain types of improper integrals whose integrand is the product of a rational function of x and sine or cosine terms.

Example

Show that $I = \int_0^\infty \frac{\cos x}{\left(1+x^2\right)} dx = \frac{\pi}{2e}$

Solution

Firstly, let re-write the integrand and its equivalent form:

$$f(x) = \frac{\cos x}{\left(1+x^2\right)} \Leftrightarrow f(z) = \frac{\cos z}{\left(1+z^2\right)}$$

Next, let us determine the poles of the integrand – if any –, we noticed that the complex function has two poles; they are:

$1+z^2 = 0 \Rightarrow z_1 = -i, z_2 = +i$, both of order two

For the same reason mentioned in case (1) of improper integrals, the poles in the half-lower part will be ignored. Next, we are now ready to evaluate the residues corresponding in the upper-half part, and it is only one pole, so we will evaluate the residue corresponding to $z_2 = +i$ only.

$$\operatorname{Re}s\left[f(z), z=i\right] = \frac{1}{(2-1)!} \lim_{z \to i} \frac{d}{dz}\left[(z-i)^2 \frac{\cos z}{(z-i)^2 (z+i)^2}\right]$$

$$\operatorname{Re}s\left[f(z), z=i\right] = \lim_{z \to i} \frac{d}{dz}\left[\frac{\cos z}{(z+i)^2}\right]$$

$$\operatorname{Re}s\left[f(z), z=i\right] = \lim_{z \to i}\left[\frac{-\sin z(z+i) - 2\cos z}{(z+i)^3}\right]$$

$$\operatorname{Re}s\left[f(z), z=i\right] = \left[\frac{-2(\cos i + i\sin i)}{(2i)^3}\right]$$

But

$$\left(\cos i + i\sin i\right) = e^{i(i)}$$

Therefore,

$$\operatorname{Re}s\left[f\left(z\right), z = i\right] = \left[\frac{-2e^{i(i)}}{\left(2i\right)^{3}}\right] = \left[\frac{e^{-1}}{4i}\right]$$

Now then the value of the given integral will be:

$$I = \oint_{C} \frac{\cos z}{\left(1+z^{2}\right)} dz = \left(2\pi i\right)\operatorname{Re}s\left(f, z_{j}\right) = \left(2\pi i\right)\left[\frac{e^{-1}}{4i}\right] = \left[\frac{\pi}{2e}\right]$$

Example

Show that $I = \int_{0}^{\infty} \left(\frac{x\sin x}{x^{2}+4}\right) dx = \frac{\pi e^{=2}}{2}$

Solution

Firstly, let us re-write the integrand and its equivalent form:

$$f\left(x\right) = \left(\frac{x\sin x}{x^{2}+4}\right) \Leftrightarrow f\left(z\right) = \left(\frac{z\sin z}{z^{2}+4}\right)$$

In this case of problem, let deal with the function:

$$f\left(x\right) = \left(\frac{x\sin x}{x^{2}+4}\right) \Leftrightarrow f\left(z\right) = \left(\frac{ze^{iz}}{z^{2}+4}\right)$$

And we will deal with the integral that takes the following form:

$$\int_{-\infty}^{\infty} \left(\frac{xe^{ix}}{x^{2}+4}\right) dx = \int_{-\infty}^{\infty} \left(\frac{ze^{iz}}{z^{2}+4}\right) dz$$

Next, let us determine the poles of the integrand – if any –, we noticed that the complex function has one pole in the upper-half plane:

$z^2+4=0 \Rightarrow z_1=-2i, z_2=+2i$, both of order one

For the same reason mentioned in case (1) of improper integrals, the poles in the half-lower part will be ignored. Next, we are now ready to evaluate the residues corresponding in the upper-half part, and it is only one pole, so we will evaluate the residue corresponding to $z_2=+2i$ only.

$$\mathrm{Res}\left[f(z), z=2i\right]=\lim_{z\to 2i}\left[(z-2i)\frac{ze^{iz}}{(z-2i)(z+2i)}\right]$$

$$\mathrm{Res}\left[f(z), z=2i\right]=\lim_{z\to 2i}\left[\frac{ze^{iz}}{(z+2i)}\right]$$

$$\mathrm{Res}\left[f(z), z=2i\right]=\left[\frac{e^{-2}}{2}\right]$$

Now then the value of the given integral will be:

$$I=\int_{-\infty}^{\infty}\left(\frac{ze^{iz}}{z^2+4}\right)dz=\oint_C\left(\frac{ze^{iz}}{z^2+4}\right)dz=(2\pi i)\mathrm{Res}(f, z=2i)$$

$$=(2\pi i)\left(\frac{e^{-2}}{2}\right)=\pi i e^{-2}$$

which is the same as:

$$I=\int_{-\infty}^{\infty}\left(\frac{x\cos x}{x^2+4}+i\frac{x\sin x}{x^2+4}\right)dx=\pi i e^{-2}$$

Equating the real parts in sides and also the imaginary parts, leads to:

$$\int_{-\infty}^{\infty}\left(\frac{x\sin x}{x^2+4}\right)dx=\pi\, e^{-2}$$

But the integrand is an even function, then:

$$\int_{-\infty}^{\infty}\left(\frac{x\sin x}{x^2+4}\right)dx=2\int_{0}^{\infty}\left(\frac{x\sin x}{x^2+4}\right)dx=\pi\, e^{-2}$$

$$2\int_{0}^{\infty}\left(\frac{x\sin x}{x^2+4}\right)dx=\pi\, e^{-2}$$

$$\Rightarrow$$

$$\int_{0}^{\infty}\left(\frac{x\sin x}{x^2+4}\right)dx=\frac{\pi\, e^{-2}}{2}$$

Example

Show that $\int_{0}^{\infty}\cos x^2\,dx=\int_{0}^{\infty}\sin x^2\,dx=\frac{\sqrt{2\pi}}{4}$

Solution

First of all, let us have a look to the graph of the function $\sin x^2$, as shown in Fig. (**5**).

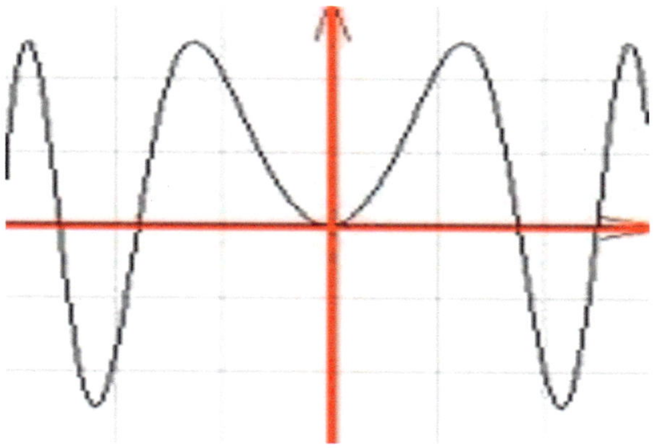

Figure 5: Graph of $\sin x^2$.

Now let us start with the function $f(z) = e^{iz^2}$ and choose the appropriate contour shown in Fig. (**6**).

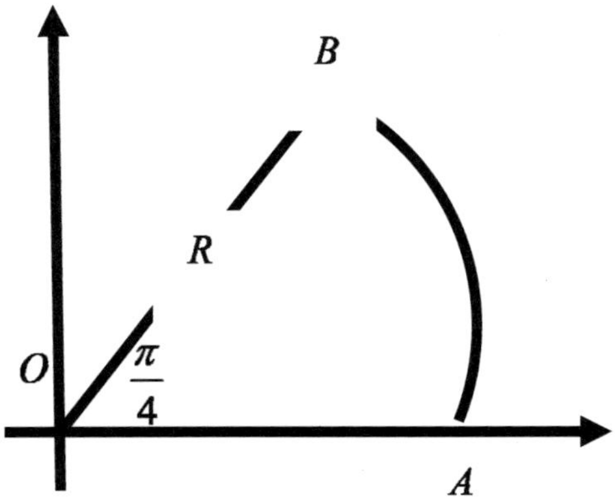

Figure 6: Contour of the function.

Since the function $f(z) = e^{iz^2}$ is analytic everywhere, and we know from Cauchy theorem of integration that the integration of analytic function around any closed contour is equal to zero.

$$\oint_C f(z)\,dz = 0$$

or

$$\oint_C e^{iz^2}\,dz = \oint_{\overrightarrow{OA}} e^{iz^2}\,dz + \oint_{\overset{\frown}{AB}} e^{iz^2}\,dz + \oint_{\overrightarrow{BO}} e^{iz^2}\,dz = 0$$

Now on the line segment $\overrightarrow{OA}$, we have:

$$z = x$$
$$dz = dx$$

and on the arc $\overset{\frown}{AB}$, we have:

$$z = \text{R}e^{i\Theta}$$
$$dz = i\,\text{R}e^{i\Theta}\,d\Theta$$

and finally, on the radius $\overrightarrow{BO}$, we have:

$$z = r\left(\cos\frac{\pi}{4} + i\sin\frac{\pi}{4}\right) = re^{i\frac{\pi}{4}}$$

$$dz = e^{i\frac{\pi}{4}}dr$$

Now the above three integrals can be re-written as follows:

$$\oint_{\overrightarrow{OA}} e^{iz^2}dz + \oint_{\overset{\frown}{AB}} e^{iz^2}dz + \oint_{\overrightarrow{BO}} e^{iz^2}dz = 0$$

$$\int_0^R e^{ix^2}dx + \int_0^R i\,\mathrm{Re}^{i\Theta}\, e^{iR^2e^{2i\Theta}}d\Theta + \int_R^0 e^{\frac{i\pi}{4}} e^{ir^2e^{\frac{i\pi}{2}}}dr = 0$$

Transposing the 2nd and 3rd integrals to the right-hand side, leads to:

$$\int_0^R e^{ix^2}dx = -\int_0^R i\,\mathrm{Re}^{i\Theta}\, e^{iR^2e^{2i\Theta}}d\Theta - \int_R^0 e^{\frac{i\pi}{4}} e^{ir^2e^{\frac{i\pi}{2}}}dr$$

We now

$$e^{ix^2} = \cos x^2 + i\sin x^2$$

Then

$$\int_0^R \left(\cos x^2 + i\sin x^2\right)dx = -\int_0^R i\,\mathrm{Re}^{i\Theta}\, e^{iR^2e^{2i\Theta}}d\Theta - \int_R^0 e^{\frac{i\pi}{4}} e^{ir^2e^{\frac{i\pi}{2}}}dr$$

Also we know

$$e^{\frac{i\pi}{2}} = i$$

Therefore:

$$\int_0^R \left(\cos x^2 + i\sin x^2\right)dx = e^{\frac{i\pi}{4}} \int_0^R e^{-r^2}\,dr - \int_0^R i\,\mathrm{Re}^{i\Theta}\, e^{iR^2 e^{2i\Theta}}\,d\Theta$$

Then

$$\int_0^R \left(\cos x^2 + i\sin x^2\right)dx = e^{\frac{i\pi}{4}} \left[\int_0^R e^{-r^2}\,dr\right] - iR\int_0^R e^{i\Theta} e^{iR^2(\cos 2\Theta + i\sin 2\Theta)}\,d\Theta$$

Now let us dealing with the 1st integral in the right-hand side of the above equation:

$$I = \left[\int_0^R e^{-r^2}\,dr\right]$$

As $R \to \infty$ the integral I will be:

$$I = \left[\int_0^R e^{-r^2}\,dr\right] = \left[\int_0^\infty e^{-r^2}\,dr\right]$$

Since

$$\Gamma\left(\frac{1}{2}\right) = 2\int_0^\infty e^{-x^2}\,dx = \sqrt{\pi}$$

Or

$$\int_0^\infty e^{-x^2}\,dx = \frac{\sqrt{\pi}}{2}$$

Now then:

$$\int_0^R \left(\cos x^2 + i\sin x^2\right)dx = \frac{\sqrt{\pi}}{2}\left(\cos\frac{\pi}{4} + i\sin\frac{\pi}{4}\right) - iR\int_0^{\frac{\pi}{4}} e^{i\Theta} e^{iR^2(\cos 2\Theta + i\sin 2\Theta)}\,d\Theta$$

Note that:

$$\frac{\sqrt{\pi}}{2}\left(\cos\frac{\pi}{4}+i\sin\frac{\pi}{4}\right)=\frac{\sqrt{2\pi}}{4}(i+1)$$

$$\int_0^R\left(\cos x^2+i\sin x^2\right)dx=\frac{\sqrt{2\pi}}{4}(i+1)\underbrace{-iR\int_0^{\frac{\pi}{4}}e^{i\Theta}e^{iR^2(\cos 2\Theta+i\sin 2\Theta)}d\Theta}_{I_1}$$

Now, let us deals with the 2nd term in the right-hand side,

$$I_1=-iR\int_0^{\frac{\pi}{4}}e^{i\Theta}e^{iR^2(\cos 2\Theta+i\sin 2\Theta)}d\Theta$$

Let us considering the absolute value of the above integral:

$$\left|-iR\int_0^{\frac{\pi}{4}}e^{i\Theta}e^{iR^2(\cos 2\Theta+i\sin 2\Theta)}d\Theta\right|\le R\int_0^{\frac{\pi}{4}}\left|e^{i\Theta}\right|\left|e^{iR^2\cos 2\Theta}\right|\left|e^{-iR^2\sin 2\Theta}\right|d\Theta$$

$$=R\int_0^{\frac{\pi}{4}}e^{-R^2\sin 2\Theta}d\Theta$$

If we change the variable of integration from Θ to α using the following substitution:

$$\alpha=2\Theta\Rightarrow d\alpha=2d\Theta$$

New limits of integration will be then:

$$\Theta=0\Leftrightarrow\alpha=0$$

$$\Theta=\frac{\pi}{4}\Leftrightarrow\alpha=\frac{\pi}{2}$$

Now then, the integral in terms of the new variable takes the following form:

$$I = R\int_0^{\frac{\pi}{4}} e^{-R^2 \sin 2\Theta} d\Theta = \frac{R}{2}\int_0^{\frac{\pi}{2}} e^{-R^2 \sin\alpha} d\alpha$$

To evaluate the last integral, we have two methods, they are:

First Method

By using the following trigonometric substitution:

$$y = \sin\alpha$$

In this case, the integral takes the following form:

$$I = \frac{R}{2}\int_0^{\frac{\pi}{2}} e^{-R^2 \sin\alpha} d\alpha = \frac{R}{2}\int_0^{1} \frac{e^{-R^2 y}}{\sqrt{1-y^2}} d\alpha$$

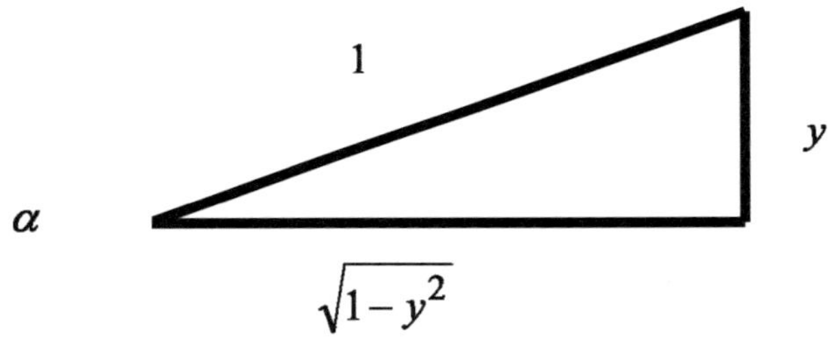

Figure 7: Right triangle.

Second Method

By intersection method, this can be carried out by assuming line passing through $\alpha = 0 \,\&\, \alpha = \frac{\pi}{2}$, then from $\alpha = 0 \to \frac{\pi}{2}$, we have $\sin\alpha \geq \frac{2\alpha}{\pi}$, then:

$$I = \frac{R}{2}\int_0^{\frac{\pi}{2}} e^{-R^2 \sin\alpha} d\alpha = \frac{R}{2}\int_0^{1} e^{\frac{-2\alpha R^2}{\pi}} d\alpha$$

$$I = \frac{-\pi}{4R}\left(e^{\frac{-2\alpha R^2}{\pi}}\right)_0^{\frac{\pi}{2}} = \frac{\pi}{4R}\left(1 - e^{-R^2}\right)$$

As $R \to \infty$

$$I = \frac{R}{2}\int_0^{\frac{\pi}{2}} e^{-R^2 \sin\alpha} d\alpha = 0$$

Then

$$\int_0^R \left(\cos x^2 + i \sin x^2\right) dx = \frac{\sqrt{2\pi}}{4}(1+i)$$

Equating real and imaginary parts in both sides, one can get:

$$\int_0^{\infty} \cos x^2 dx = \int_0^{\infty} \sin x^2 dx = \frac{\sqrt{2\pi}}{4}$$

5.5. GENERAL SOLVED EXAMPLES

Example

Find the residue for $f(z) = \dfrac{2z+3}{z^2-4}$

Solution

First of all, let us determine the poles of the given complex-valued function:

$$f(z) = \frac{2z+3}{z^2-4} = \frac{2z+3}{(z-2)(z+2)} \qquad \text{(A-1)}$$

Now we have two poles, they are:

$z = 2$, **order 1** (A-2)

$z = -2$, **order 1** (A-3)

Next, let us evaluate the residue corresponding to the pole defined by equation (A-2), as follows:

$$\operatorname{Res}(f(z), z = 2) = \lim_{z \to 2} \left[\underbrace{(z-2)}_{\textbf{from which we found the pole}} \frac{2z+3}{(z-2)(z+2)} \right]$$
$$= \frac{7}{4} \tag{A-4}$$

Next, let us evaluate the residue corresponding to the pole defined by equation (A-3), as follows:

$$\operatorname{Res}(f(z), z = -2) = \lim_{z \to -2} \left[\underbrace{(z+2)}_{\text{from which we found the pole}} \frac{2z+3}{(z-2)(z+2)} \right] \tag{A-5}$$
$$= \frac{1}{4}$$

Example

Find the residue for $f(z) = \dfrac{e^z}{(z-2)^3}$

Solution

First of all, let us determine the poles of the given complex-valued function:

$$f(z) = \frac{e^z}{(z-2)^3} \tag{A-1}$$

Now we have only one pole, it will be:

$$z = 2, \text{ order } 3 \tag{A-2}$$

Next, let us evaluate the residue corresponding to the pole defined by equation (A-2), as follows:

$$\operatorname{Res}(f, z=2) = \frac{1}{(2)!} \lim_{z \to 2} \left[\frac{d^2}{dz^2} \left(\underbrace{(z-2)^3}_{\text{from which we found the pole}} \frac{2z+3}{(z-2)^3} \right) \right] \quad \text{(A-3)}$$

$$= \frac{e^2}{2}$$

Example

Evaluate the following integral $I = \oint_C \frac{z^2}{(z+1)(z+3)} dz$, note that the path C encloses all poles.

Solution

The first step as we did before so many times is to determine the poles, as follows:

$$f(z) = \frac{z^2}{(z+1)(z+3)}$$

$$\Rightarrow \quad \text{(A-1)}$$

Poles are: $z = -1$ & $z = -3$ Both of order 1

As long as all the poles are interior of the given path, so we should evaluate the residue corresponding to each pole:

Residue Corresponding to $z = -1$

$$\operatorname{Res}(f(z), z=-1) = \lim_{z \to -1} \left[\underbrace{(z+1)}_{\text{from which we found the pole}} \frac{z^2}{(z+1)(z+3)} \right] \quad \text{(A-2)}$$

$$= \frac{1}{2}$$

Residue Corresponding to $z = -3$

$$\operatorname{Res}\left(f(z), z=-3\right) = \lim_{z\to -3}\left[\underbrace{(z+3)}_{\text{from which we found the pole}} \frac{z^2}{(z+1)(z+3)}\right] = -\frac{9}{2} \quad \text{(A-3)}$$

We are now ready to evaluate the integral, as follows:

$$\begin{aligned} I &= (2\pi i)\sum_{j=1}^{2}\operatorname{Res}\left(f(z), z_j\right) \\ &= (2\pi i)\left[\frac{1}{2}-\frac{9}{2}\right] \\ &= -(8\pi i) \end{aligned} \quad \text{(A-4)}$$

Example

Prove that:

$$I = \oint_C \frac{ze^{zt}}{\left(z^2+1\right)^2}dz = \frac{1}{2}\sin t, \text{ note that the path C encloses } z = \pm i.$$

Solution

The first step as we did before so many times is to determine the poles, as follows:

$$f(z) = \frac{ze^{zt}}{\left(z^2+1\right)^2}$$

$$\Rightarrow \quad \text{(A-1)}$$

Poles are: $z = \pm i$ order 2

As long as all the poles are interior of the given path, so we should evaluate the residue corresponding to each pole:

Residue Corresponding to $z = i$

$$\begin{aligned}\operatorname{Res}(f, z=i) &= \lim_{z\to i}\frac{d}{dz}\left[\underbrace{(z+1)^2}_{\text{from which we found the pole}}\frac{ze^{zt}}{(z-i)^2(z+i)^2}\right] \\ &= -\frac{1}{4}ite^{it}\end{aligned} \tag{A-2}$$

Residue Corresponding to $z = -i$

$$\begin{aligned}\operatorname{Res}(f, z=-i) &= \lim_{z\to -i}\frac{d}{dz}\left[(z+1)^2\frac{ze^{zt}}{(z-i)^2(z+i)^2}\right] \\ &= \frac{1}{4}ite^{-it}\end{aligned} \tag{A-3}$$

We are now ready to evaluate the integral, as follows:

$$\begin{aligned}I &= (2\pi i)\sum_{j=1}^{2}\operatorname{Res}(f(z), z_j) \\ &= (2\pi i)\left[-\frac{1}{4}ite^{it} + \frac{1}{4}ite^{-it}\right] \\ &= (\pi it)\sin t\end{aligned} \tag{A-4}$$

Example

Integrate $I = \oint_C \frac{z}{(z^2-1)}dz$, around the path $|z| = 4$.

Solution

The first step as we did before so many times is to determine the poles, as follows:

$$f(z) = \frac{z}{(z^2 - 1)}$$

$$\Rightarrow \tag{A-1}$$

Poles are: $z = \pm 1$ order 1

As long as all the poles are interior of the given path, so we should evaluate the residue corresponding to each pole:

Residue Corresponding to $z = 1$

$$\begin{aligned} \operatorname{Res}(f, z=1) &= \lim_{z \to 1} \left[\underbrace{(z-1)}_{\text{from which we found the pole}} \frac{z}{(z-1)(z+1)} \right] \\ &= \frac{1}{2} \end{aligned} \tag{A-2}$$

Residue Corresponding to $z = -1$

$$\begin{aligned} \operatorname{Res}(f, z=-1) &= \lim_{z \to -1} \left[\underbrace{(z+1)}_{\text{from which we found the pole}} \frac{z}{(z-1)(z+1)} \right] \\ &= -\frac{1}{2} \end{aligned} \tag{A-3}$$

We are now ready to evaluate the integral, as follows:

$$\begin{aligned} I &= (2\pi i) \sum_{j=1}^{2} \operatorname{Res}(f(z), z_j) \\ &= (2\pi i) \left[-\frac{1}{2} + \frac{1}{2} \right] = 0 \end{aligned} \tag{A-4}$$

This result agrees well with the theorem, that integration over a closed path equals zero.

Example

Integrate $I = \oint_C \tan z dz$, around the path $|z| = 2$.

Solution

The first step as we did before so many times is to determine the poles, as follows:

$$f(z) = \tan z dz = \frac{\sin z}{\cos z}$$

$$\Rightarrow \tag{A-1}$$

Poles are: $z = -\frac{\pi}{2}, z = +\frac{\pi}{2}$ both of order 1

We know if we have $P(z) = \frac{f(z)}{g(z)}$ and $f(z_0) \neq 0$ but $g(z_0) = 0$, then:

$$\operatorname{Res}\left[P(z), z_0\right] = \frac{f(z_0)}{g'(z_0)} \tag{A-2}$$

Now we are ready to evaluate the residue corresponding to each pole.

$$\operatorname{Res}\left(f(z), z = -\frac{\pi}{2}\right) = \left[\frac{\sin\left(-\frac{\pi}{2}\right)}{-\sin\left(-\frac{\pi}{2}\right)}\right] = -1 \tag{A-3}$$

$$\operatorname{Res}\left(f(z), z = \frac{\pi}{2}\right) = \left[\frac{\sin\left(\frac{\pi}{2}\right)}{-\sin\left(\frac{\pi}{2}\right)}\right] = -1 \tag{A-4}$$

$$\begin{aligned} I &= (2\pi i)\sum_{j=1}^{2} \operatorname{Res}\left(f(z), z_j\right) \\ &= (2\pi i)[-1-1] \\ &= -(4\pi i) \end{aligned} \tag{A-5}$$

Chapter 6: Transformation and Conformal Mapping

CHAPTER 6

Transformation and Conformal Mapping

Abstract: In complex analysis, a function preserves angles if and only if it is analytic or anti-analytic the conjugate of an analytic function. In the present chapter, we will deal with the transformation and conformal mapping, let us start this chapter, by introducing the meaning of the word "**Conformal**", then introducing some basic definitions for both mapping and transformation. **Question**: what does it mean by the word conformal? The answer is quite simple, the word conformal means preserving angle of the shape after transformation at a certain point.

Keywords: Bilinear transformation, inverse transformation, linear transformation, power transformation, transformations.

6.1. DEFINITION: MAPPING

Mapping is the process of sending points of z-plane onto points of the w-plane **[36]**.

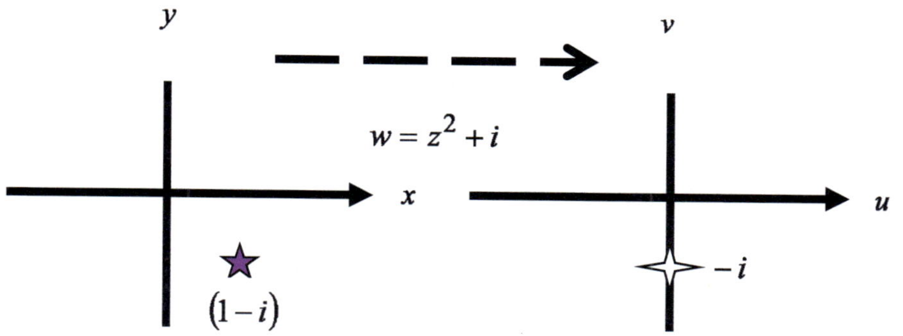

Figure 1: Mapping from z-plane onto w-plane.

Referring to Fig. (**1**), and let us check the definition by the mapping that is used to transfer the point $(1-i)$ in the z-plane onto the point $(-i)$ in the w-plane.

The given mapping:

$$w = z^2 + i$$

Now, substituting by the point $(1-i)$ in the z-plane in the given mapping, leads to:

$$w = (1-i)^2 + i = -i$$

As we see, we obtained the corresponding point in the w-plane.

6.2. DEFINITION: TRANSFORMATION

Transformation is the process of transferring a figure from z-plane to the w-plane.

Let us take an example of a transferring the square given in Fig. (**2**) in the z-plane to the corresponding one in the w-plane.

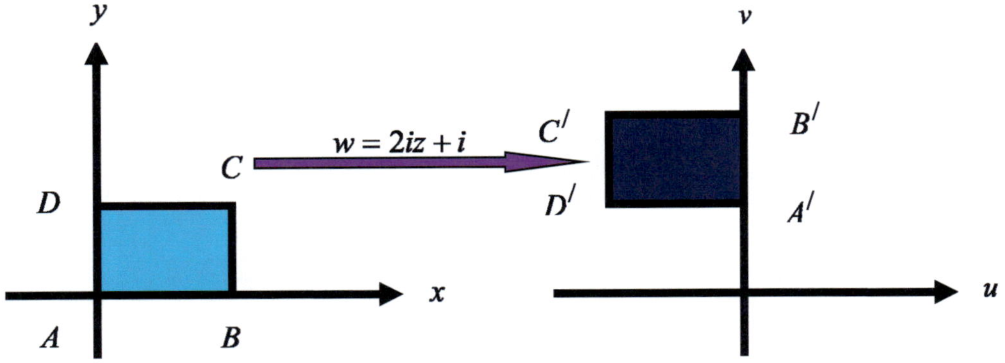

Figure 2: Mapping from z-plane onto w-plane.

The vertices of the square in the z-plane shown in Fig. (**2**) –left–are as follows:

$$A = (0,0)$$

$$B = (1,0)$$

$$C = (1,1+i)$$

$$D = (0,i)$$

The corresponding vertices of the square in the w-plane shown under the transformation $w = 2iz + i$ in Fig. (**2**) –right– are as follows:

$$A^{/} \Leftrightarrow w = 2iz + i = 2i(0+0i) + i = i$$

$$B' \Leftrightarrow w = 2iz + i = 2i(1 + 0i) + i = 3i$$

$$C' \Leftrightarrow w = 2iz + i = 2i(1 + [1 + i0]i) + i = 3i - 2$$

$$D' \Leftrightarrow w = 2iz + i = 2i(0 + [i]i) + i = -2 + i$$

There are two types of mapping, they are:

1. One-to-One Mapping

A mapping $w = f(z)$ under which each point in the z-plane has corresponding only one point in the w-plane, see Fig. (**3**).

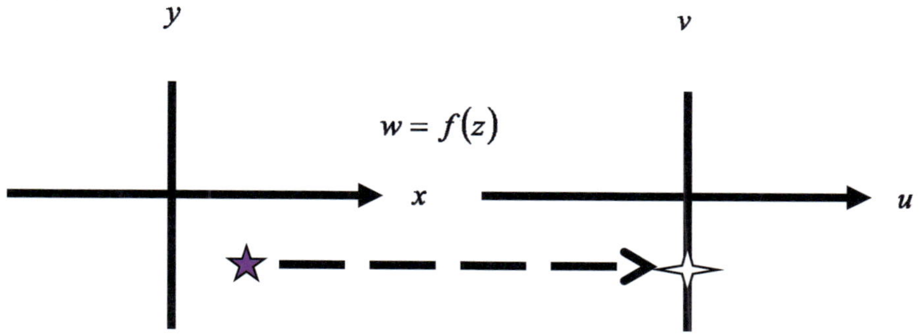

Figure 3: One-to-one mapping from z-plane onto w-plane.

2. Many-to-One Mapping

In this type, there are many points in the z-plane and only one corresponding point in the w-plane, see Fig. (**4**).

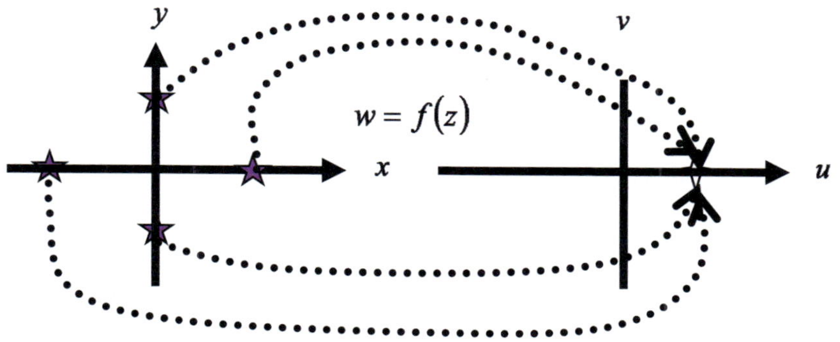

Figure 4: Many-to-one mapping from z-plane onto w-plane.

6.3. LINEAR TRANSFORMATION

The linear transformation takes the form:

$$w = f(z) = az + b \tag{1}$$

In equation (1), both a and b are constants, whatever they are real or complex. As we see the linear transformation consists of two major parts, next we will deal with each part separately.

First Part: Rotation Stretching

This part of linear transformation takes the form:

$$\xi = az \tag{2}$$

As we see from its name, rotation stretching, the reason for this terminology becomes apparent when one examines the relation. From this part of translation, one can deduce the following mathematical form:

$$|\xi| = |a||z| \tag{3}$$

And

$$\arg(\xi) = \arg(a) + \arg(z) \tag{4}$$

Note that: The part $|a|$ in equation (3) referred as magnification, while $|z|$, and referred as rotation. This means that the linear transformation consists of two parts, magnification and rotation.

There are three different cases encountered when dealing with linear transformation, they are:

Case (1): No rotation & No Translation

If $\arg(a) = 0 \quad \& \quad |a| = 1 \Rightarrow \xi = z$

This means that there are no rotation and translation.

Case (2): Pure Rotation & No Translation

If $\arg(a) \neq 0$ & $|a| = 1 \Rightarrow \xi = z$

This means that there is pure rotation but no translation.

Case (3): Magnification (Stretching or Shrinkage)

If $\arg(a) = 0$

Herein, we have two different cases of magnification depends mainly on the value of..will be as follows:

$$\begin{array}{lcl} |a| > 1 & \Leftrightarrow & \text{Stretching} \\ |a| < 1 & \Leftrightarrow & \text{Shrinkage} \end{array} \qquad (5)$$

We noticed that such a mapping preserves singularity, since it rotates every point through the same angle, namely $\arg(a)$ by the same factor – Stretching or Shrinkage $-|a|$.

First Part: Translation

The linear transformation is now then, takes the following new form:

$$w = \xi + b \qquad (6)$$

Equation (6) is then now is called the translation, and has property that it shifts or translates every point ξ by the constant vector b. This transformation involves neither rotation nor stretching.

Example

Under the following function $w = f(z) = 2iz + (1+i)$ transform each of the following:

1. The point $P:(1+2i)$

2. The line segment $S : \arg(z) = \frac{\pi}{4}, 1 < |z| < 2$

3. Circular arc $A : |z| = 2, \frac{3\pi}{4} \le \arg(z) \le \frac{3\pi}{2}$

4. Vertical line $L : R(z) = \frac{1}{2}$

Solution

Re-write the given function as follows:

$$w = f(z) = \underbrace{2i}_{=a} z + \underbrace{(1+i)}_{b}$$

Therefore

$$a = 2i$$

$$b = (1+i)$$

By existence of a and b this means that there are both rotation and translation, and let us now start solution each part separately, but before of that, let us evaluate $\arg(a)$ and $|z|$ to check existence of rotation and signification. Also we must know in advance that as long as b has a value $(1+i)$, this means that there will be translation.

$$\arg(a) = \tan^{-1}\left(\frac{I}{R}\right) = \tan^{-1}\left(\frac{2}{0}\right) = \tan^{-1}(\infty) = \frac{\pi}{2}$$

Case (1): Point $P : (1+2i)$

The point $P : (1+2i)$, see Fig. (**5**).

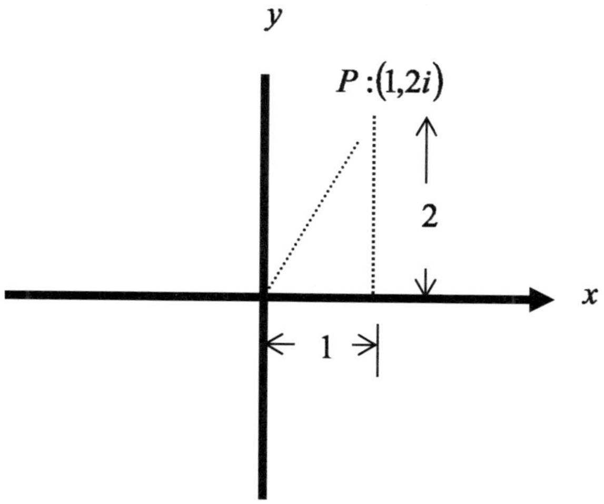

Figure 5: The original location of the point P.

Dealing with Rotation

This point will be rotated through an angle equals $\arg(a)=\frac{\pi}{2}$, therefore after rotation the old point becomes new rotated one called P_{Rotated} as shown in Fig. (**6**).

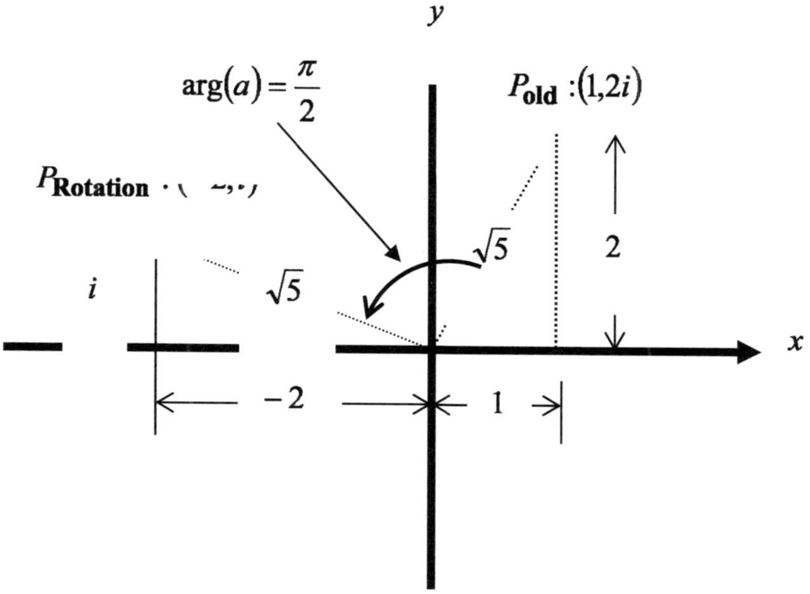

Figure 6: The new location of the point P_{old} after rotation $P_{\text{Rotation}}:(-2,i)$.

Dealing with Magnification

The point after rotation will be magnified by the value $|a| = |2i| = 2$, therefore after rotation and magnification the old point becomes P_{Rotated} + magnification as shown in Fig. (**7**).

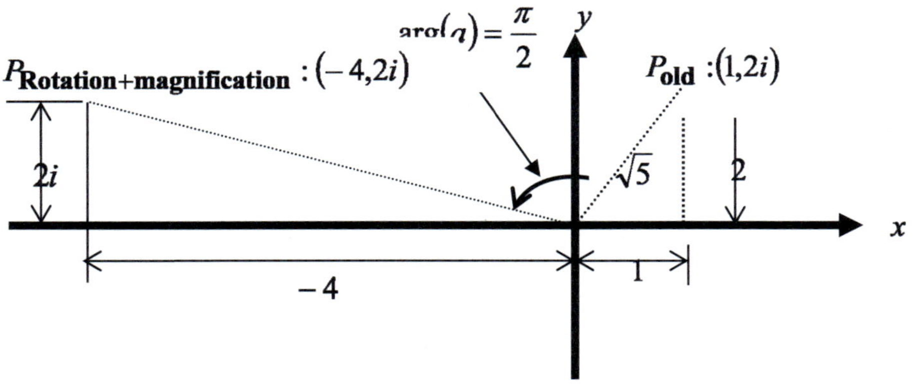

Figure 7: The location of the point P_{old} after rotation and magnification $P_{\text{Rotation+magnification}} : (-4, 2i)$.

Dealing with Translation

The point after rotation, magnification will be translated by the value $b = (1 + i)$.

Therefore the old point becomes P_{Rotated} + magnification + translation as shown in Fig. (**8**).

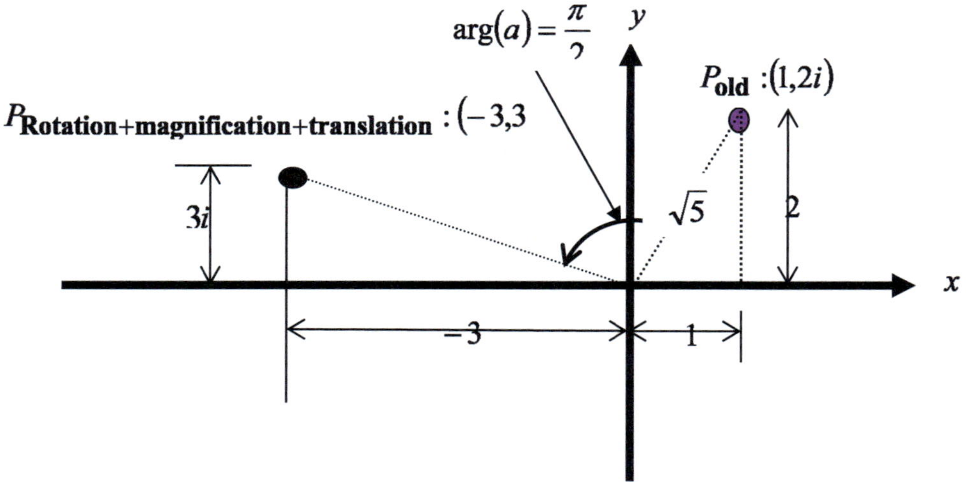

Figure 8: The location of the point P_{old} after rotation and magnification $P_{\text{Rotation+magnification+translation}} : (-3, 3i)$.

Case (2): Line Segment $S:\arg(z)=\frac{\pi}{4},\ 1<|z|<2$

The original line segment is shown in Fig. (**9**).

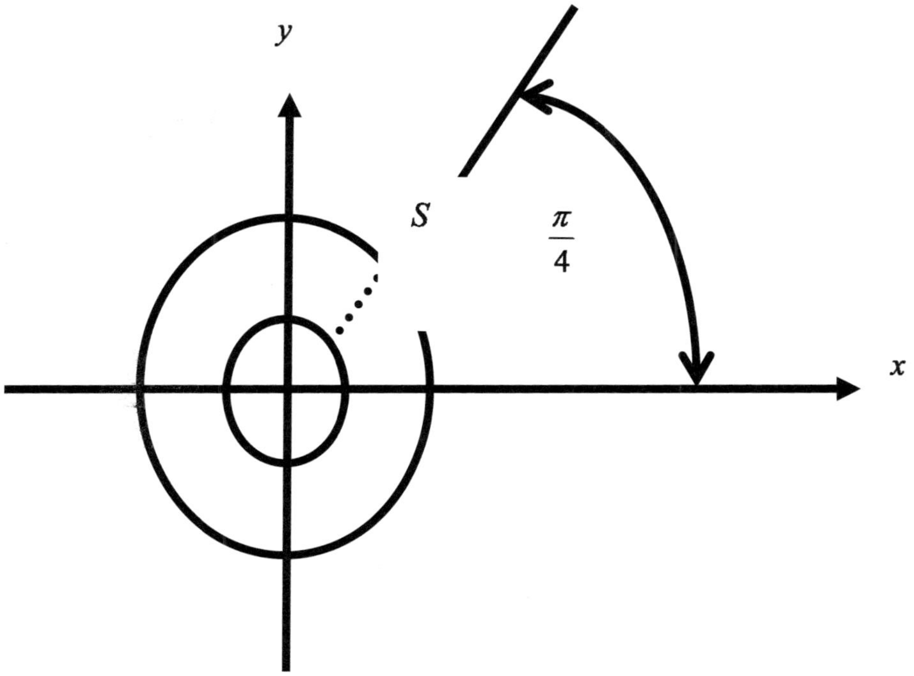

Figure 9: The original location of the line segment.

From Fig. (**9**), we noticed that the original line segment makes an angle $\frac{\pi}{4}$ with the horizontal axis in the z-plane, now we are ready to start dealing with the first process, 'Rotation'.

Dealing with Rotation

The line segment originally makes an angle $\frac{\pi}{4}$ with the horizontal axis and because $\arg(a)=\frac{\pi}{2}$, therefore, it will be rotated through an angle equals $\arg(a)=\frac{\pi}{2}$ in addition to the original angle of inclination, therefore after rotation the old line segment becomes new rotated one called S_{Rotated} as shown in Fig. (**10**).

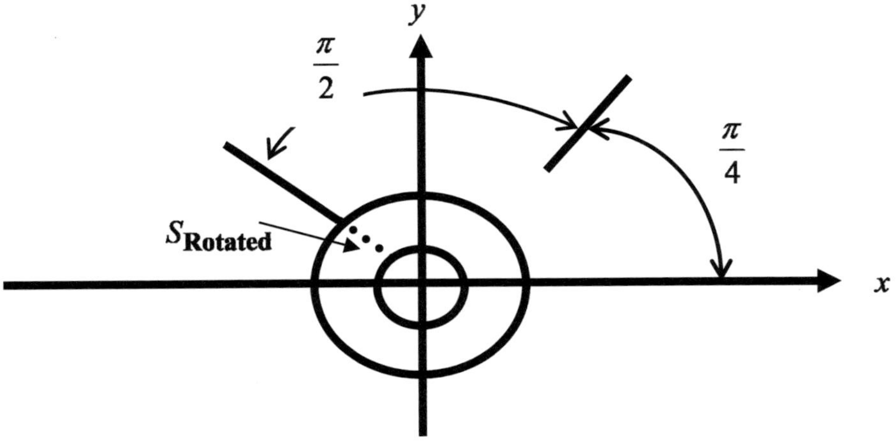

Figure 10: The new location of the line segment S_{old} after rotation S_{Rotated}.

Dealing with Magnification

The line segment after rotation will be magnified by the value $|a| = |2i| = 2$, therefore after rotation and magnification the line segment becomes S_{Rotated} + magnification as shown in Fig. (**11**).

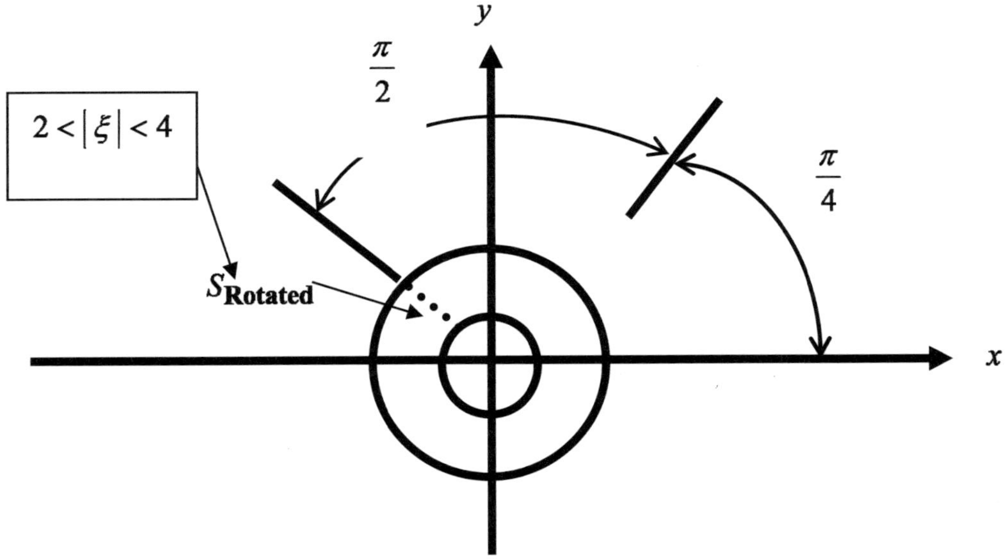

Figure 11: The location of the old line segment S_{old} after rotation and magnification $S_{\text{Rotation+magnification}}$.

Dealing with Translation

Dealing with translation herein should be taken carefully, and this can be done by re-writing the equations of the line segment after the previous rotation and magnification, they are:

$$S_{\text{Rotation+magnification}} : \arg(z) = \frac{3\pi}{4},\ 2 < |z| < 4$$

Therefore the line segment becomes $S_{\text{Rotated}} + \text{magnification} + \text{translation}$,will have the same angle $\frac{3\pi}{4}$, but the center of the two circles will move at $(1,i)$, that is $2 < |w-(1+i)| < 4$ as shown in Fig. (**12**).

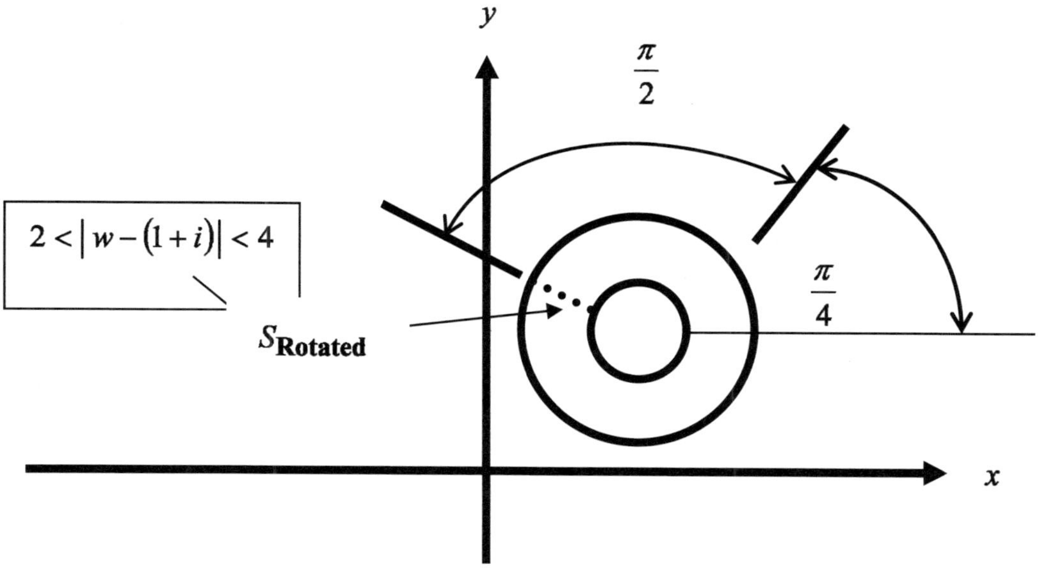

Figure 12: The location of the line segment S_{old} after rotation and magnification $S_{\text{Rotation+magnification+translation}}$.

Case (3): Circular Arc $A : |z| = 2,\ \frac{3\pi}{4} \le \arg(z) \le \frac{3\pi}{2}$

The original circular arc is shown in Fig. (**13**).

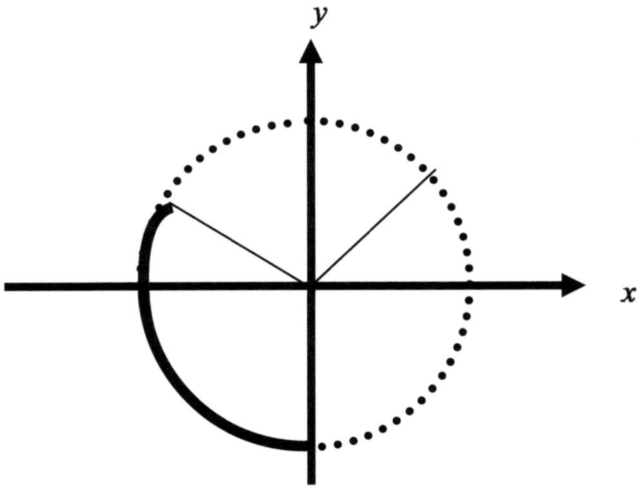

Figure 13: The original location of the circular arc.

From Fig. (**13**), we noticed that the original circular arc starts with an angle $\frac{\pi}{4}+\frac{\pi}{2}=\frac{3\pi}{4}$ with the horizontal axis in the z-plane and ended at an angle $\frac{3\pi}{2}$, now we are ready to start dealing with the first process, 'Rotation'.

Dealing with Rotation

The circular arc will be rotated through an angle equals $\arg(a)=\frac{\pi}{2}$ in addition to the original angles, therefore after rotation the old circular arc becomes new rotated one called A_{Rotated} started at an angle $\frac{\pi}{4}+\frac{\pi}{2}+\frac{\pi}{2}=\frac{5\pi}{4}$ and ended at an angle 2π as shown in Fig. (**14**).

Dealing with Magnification

The circular arc after rotation will be magnified by the value $|a|=|2i|=2$, therefore after rotation and magnification the circular arc will be the same as Fig. (**14**) but with radius equals 4 units.

Dealing with Translation

Dealing with translation herein should be taken carefully, and this can be done by re-writing the equations of the circular arc after the previous rotation and magnification, they are:

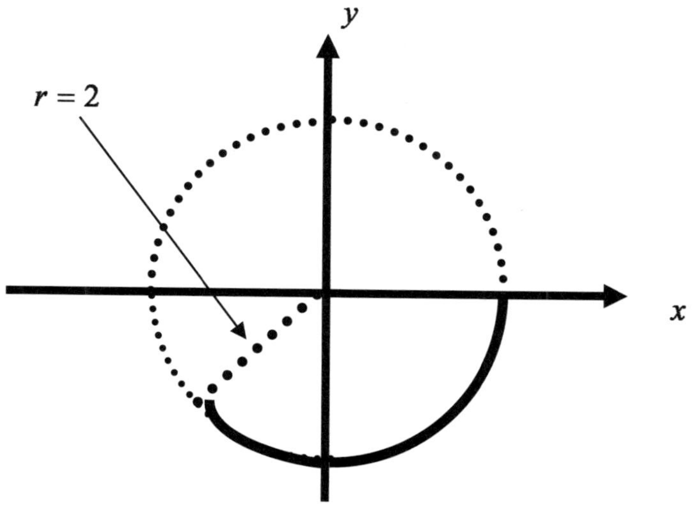

Figure 14: The new location of the circular arc A_{old} after rotation S_{Rotated}.

$$A_{\text{Rotation+magnification}} : \frac{5\pi}{4} \leq \arg(\xi) \leq 2\pi, |z| = 4$$

Therefore the circular arc becomes A_{Rotated} + magnification + translation will have a center that moves to $(1, i)$, that is $|w - (1 + i)| = 4$ as shown in Fig. (**15**).

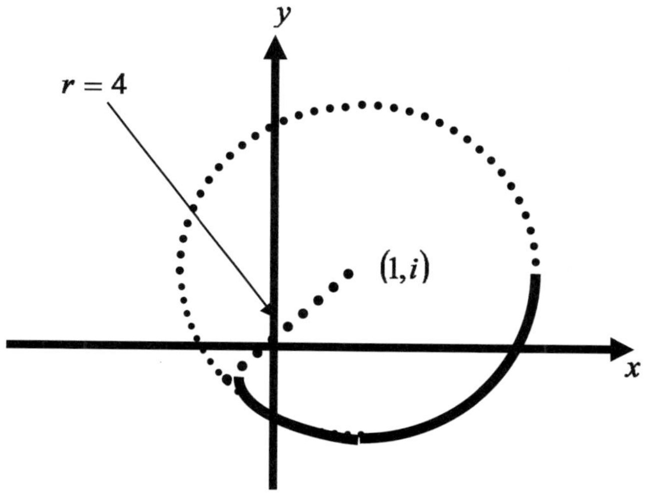

Figure 15: The location of the circular arc A_{old} after rotation and magnification $A_{\textbf{Rotation+magnification+translation}}$.

Case (4): Vertical line $L : R(z) = \frac{1}{2}$

We will leave the solution details to the reader and the final answers will be written down here:

1. Rotation leads to $L_{\textbf{Rotation}} : I(\xi) = \frac{1}{2}$

2. Magnification leads to $L_{\text{Rotation+magnification}} : I(\xi) = 1$

3. Translation leads to $L_{\text{Rotation+magnification+translation}} : I(w) = 2$

The Power Transformation

The power transformation takes the following form:

$$w = f(z) = z^n, \quad n = \text{integer} \geq 2 \tag{7}$$

Expressing equation (7) in a polar form, leads to:

$$w = r^n (\cos nt + i \sin nt) \tag{8}$$

We see readily that if:

$$|z| = r \quad \& \quad \arg(z) = t \tag{9}$$

Then:

$$|w| = r^n \quad \& \quad \arg(w) = nt \tag{10}$$

In words:

The power transformation maps a point z with modulus r and argument t onto a point with modulus r^n and argument nt .

Example

Consider the function $w = f(z) = z^2$, it is required to map the sector defined by $x = \frac{1}{2}, x = 1, y = \frac{1}{2}, y = 1$ in z-plane into w-plane under the above transformation.

Solution

First, let us re-write the function $w = f(z) = z^2$ in a more simple form as follows:

$$w = u + iv = (x + iy)^2$$

$$u + iv = x^2 - y^2 + 2ixy$$

Equating real and imaginary parts in both sides, leads to:

$$u = x^2 - y^2 \qquad \text{(Ex-1)}$$

$$v = 2xy \qquad \text{(Ex-2)}$$

Now, let us do the following analysis corresponds to the four different cases of x and y given above:

$$\text{For } x = \frac{1}{2}$$

Substituting into equations (Ex-1) and (Ex-2), respectively, leads to:

$$u = \left(\frac{1}{2}\right)^2 - y^2 \Rightarrow u = \left(\frac{1}{4}\right) - y^2$$

$$u = \left(\frac{1}{4}\right) - y^2 \Leftrightarrow x = \frac{1}{2} \qquad \text{(Ex-3)}$$

$$v = 2\left(\frac{1}{2}\right)y \Rightarrow v = y$$

$$v = y \Leftrightarrow x = \frac{1}{2} \qquad \text{(Ex-4)}$$

Combining equations (Ex-3) and (Ex-4), leads to:

$$u = \frac{1}{4} - v^2 \tag{A}$$

For $x = 1$

Substituting into equations (Ex-1) and (Ex-2), respectively, leads to:

$$u = (1)^2 - y^2 \Rightarrow u = 1 - y^2$$

$$u = u = 1 - y^2 \Leftrightarrow x = 1 \tag{Ex-5}$$

$$v = 2(1)y \Rightarrow v = 2y$$

$$v = 2y \Leftrightarrow x = 1 \tag{Ex-6}$$

Combining equations (Ex-5) and (Ex-6), leads to:

$$v^2 = 4(1 - u) \tag{B}$$

For $y = \frac{1}{2}$

Substituting into equations (Ex-1) and (Ex-2), respectively, leads to:

$$u = x^2 - \left(\frac{1}{2}\right)^2 \Rightarrow u = x^2 - \frac{1}{4}$$

$$u = x^2 - \frac{1}{4} \Leftrightarrow x = \frac{1}{2} \tag{Ex-7}$$

$$v = 2x\left(\frac{1}{2}\right) \Rightarrow v = x$$

$$v = x \Leftrightarrow x = \frac{1}{2} \tag{Ex-8}$$

Combining equations (Ex-7) and (Ex-8), leads to:

$$u = v^2 - \frac{1}{4} \tag{C}$$

For $y = 1$

Substituting into equations (Ex-1) and (Ex-2), respectively, leads to:

$$u = x^2 - (1)^2 \Rightarrow u = x^2 - 1$$

$$u = x^2 - 1 \Leftrightarrow y = 1 \tag{Ex-9}$$

$$v = 2x(1) \Rightarrow v = 2x$$

$$v = 2x \Leftrightarrow y = 1 \tag{Ex-10}$$

Combining equations (Ex-9) and (Ex-10), leads to:

$$u = \frac{v^2}{4} - 1 \tag{D}$$

Next, is to draw the region in both z-plane and w-plane, respectively, as shown in Fig. (**16**).

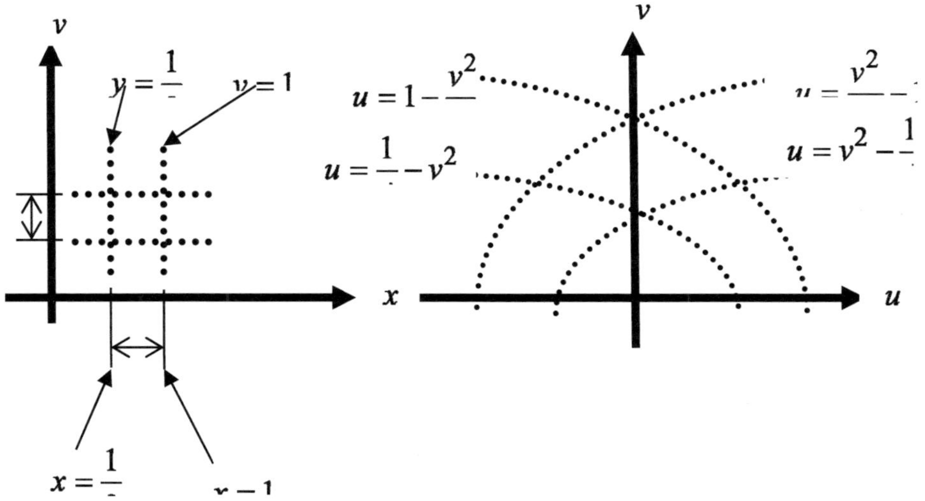

Figure 16: Region in z-plane (Left), the corresponding in w-plane (Right).

Example

Find the image of areas or functions under the transformation $w = f(z) = z^2$:

1. The straight line $y = 1 - x$
2. The line segment $1 < R(z) < 2$
3. The function $1 < |z| < 2,\ \arg(z) < \frac{\pi}{2}$

Solution

First as done before, the given function $w = f(z) = z^2$ will take the following form:

$$u = x^2 - y^2 \qquad \text{(Ex-1)}$$

$$v = 2xy \qquad \text{(Ex-2)}$$

These two equations will be used in the three parts given in the present example.

Case (1): The Straight Line..

Let us substitute into equations (Ex-1) and (Ex-2) by the given equation of the straight line, respectively leads to:

$$u = x^2 - (1-x)^2$$

$$\Rightarrow$$

$$u = (2x-1) \qquad \text{(A)}$$

And

$$v = 2x(1-x) = 2x - 2x^2$$

$$\Rightarrow$$

$$v = 2x - 2x^2 \qquad \text{(B)}$$

Combining equations (A) and (B), as follows:

From equation (A), one can get:

$$x = \left(\frac{u+1}{2}\right) \tag{C}$$

Substituting equation (c) into equation (B), leads to:

$$v = \frac{1}{2}\left(1-u^2\right) \tag{D}$$

Next, is to draw the region in both z-plane and w-plane, respectively, as shown in Fig. (**17**).

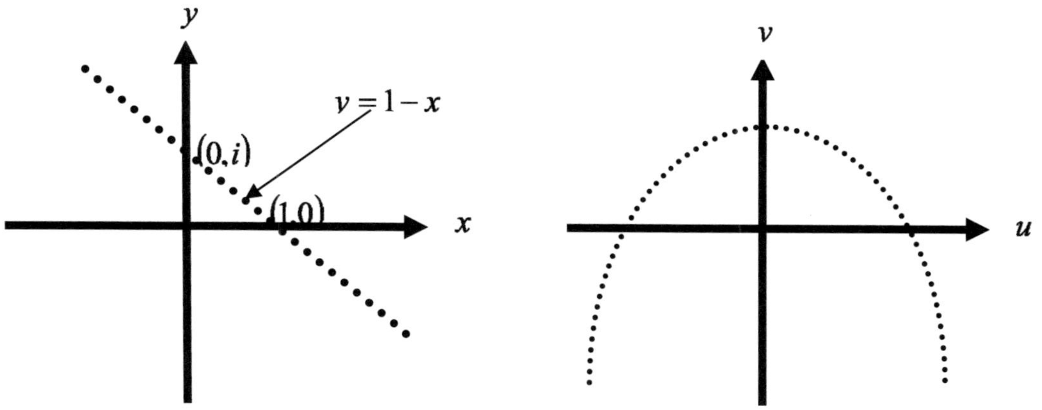

Figure 17: Region in z-plane (Left), the corresponding in w-plane (Right).

Case (2): The Line Segment $1 < R(z) < 2$

Let us substitute into equations (Ex-1) and (Ex-2) by the given equation of the line segment, respectively leads to:

For $x > 1$

$$u = 1 - y^2$$
$$v = 2y$$

$$\Rightarrow$$

..

For $x < 2$

$$u = 4 - y^2$$
$$v = 4y$$

$$\Rightarrow$$

$$u = 4 - \frac{v^2}{16}$$

Next, is to draw the region in both z-plane and w-plane, respectively, as shown in Fig. (**18**).

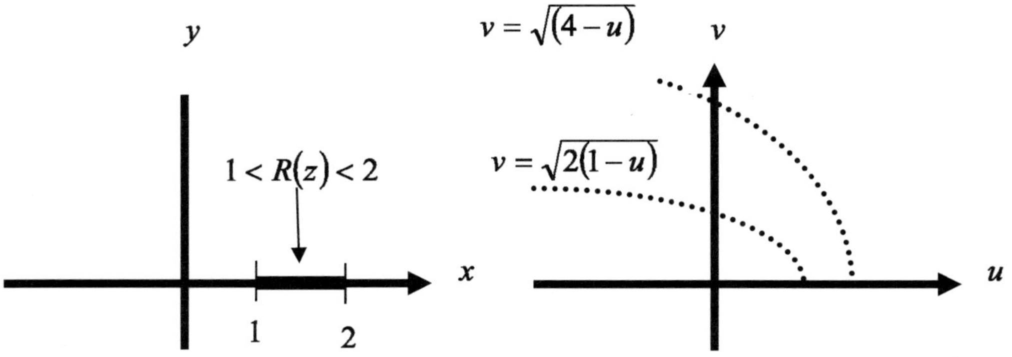

Figure 18: Region in z-plane (Left), the corresponding in w-plane (Right).

Case (2): The Function $1 < |z| < 2,\ \arg(z) < \frac{\pi}{2}$

Let us substitute into equations (Ex-1) and (Ex-2) by the given equation of the function, respectively leads to:

For $|z| > 1 \Leftrightarrow x^2 + y^2 > 1$

$$x^2 > 1 - y^2$$

Then

$$u = 1 - 2y^2$$

$$y^2 = \frac{1-u}{2} \tag{A}$$

And

$$v = 2y\sqrt{1-y^2} \tag{B}$$

Combining equations (A) and (B), leads to:

$$v = 2\left[\sqrt{\frac{1-u}{2}}\right]\left[\sqrt{1-\frac{1-u}{2}}\right] = \sqrt{1-u^2}$$

Then, we get:

$$v = \sqrt{1-u^2} \tag{C}$$

The same procedure can be carried out for $|z| < 2$. Next, is to draw the region in both z-plane and w-plane, respectively, as shown in Fig. (**19**).

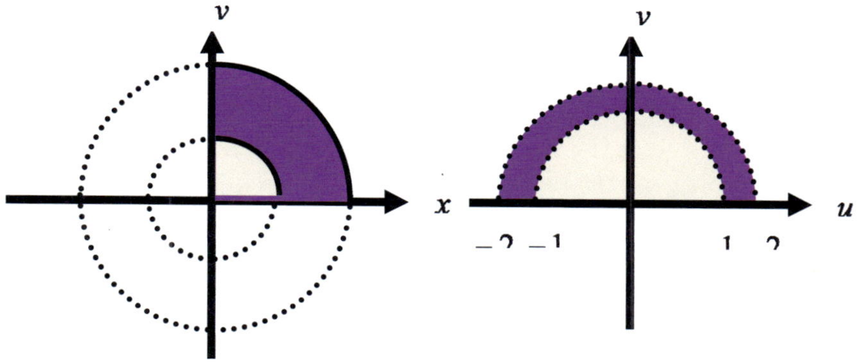

Figure 19: Region in z-plane (Left), the corresponding in w-plane (Right).

6.4. INVERSE (RECIPROCAL) TRANSFORMATION

The inverse transformation is defined as:

$$w = f(z) = \frac{1}{z} \tag{11}$$

This type of transformation has the property of mapping lines or circles in the z-plane onto lines or circles in the w-plane.

In the following example, we will discuss the property of the reciprocal (Inverse) defined above **[37]**.

Example

Find the image of the following various curves under the inverse transformation $w = f(z) = \frac{1}{z}$:

1. Vertical line $x = 1$
2. Circle $|z - 1| = 1$
3. Circle $|z| = 2$

Solution

Before dealing with the three cases, let us do some mathematical manipulation for the transformation given:

$$w = f(z) = \frac{1}{z}$$

$$u + iv = \frac{1}{x + iy}$$

Making use of the fundamentals of complex analysis, one can get the following final form:

$$u = \frac{x}{x^2 + y^2} \qquad \text{(A)}$$

$$v = -\left(\frac{y}{x^2 + y^2}\right) \qquad \text{(B)}$$

Now, we are ready to start dealing with the three cases above given in the problem.

Case (1): Vertical Line $x = 1$

Substituting the equation of the vertical straight line into equations (A) and (B), simplify leads to:

$$u = \frac{1}{1+y^2} \tag{A-1}$$

$$v = -\left(\frac{y}{1+y^2}\right) \tag{B-1}$$

Dealing with equation (A-1), yields:

$$y = \sqrt{\left(\frac{1}{u} - 1\right)} \tag{A-2}$$

Substituting from equation (A-2) into equation (B-1), yields:

$$v = -u\left[\sqrt{\left(\frac{1}{u} - 1\right)}\right] \tag{B-2}$$

Squaring both sides of equation (B-2), one can get the following:

$$u^2 + v^2 = u \tag{B-3}$$

Complete the square in equation (B-3), simplify leads to:

$$\left(u - \frac{1}{2}\right)^2 + v^2 = \frac{1}{4} \tag{B-4}$$

Equation (B-4) can then be re-written in a more suitable form as follows:

$$\left|w - \frac{1}{2}\right| = \frac{1}{2} \tag{B-5}$$

Next, is to draw the region in both z-plane and w-plane, respectively, as shown in Fig. (**20**).

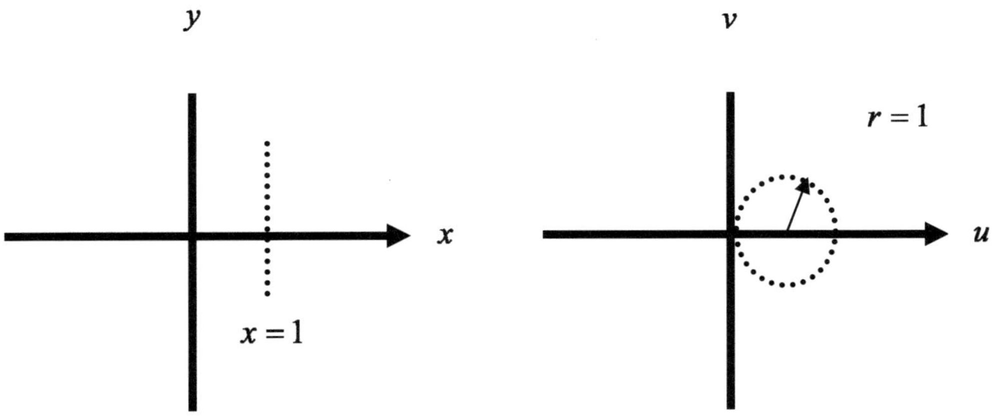

Figure 20: Vertical straight line in z-plane (Left), the corresponding is a circle in w-plane (Right).

Case (2): The Circle $|z-1|=1$

First, let us deals with the equation of the given circle as follows:

$$|z-1|=1 \tag{C-1}$$

$$(x-1)^2+y^2=1 \tag{C-2}$$

$$x^2+y^2=2x \tag{C-3}$$

Making use of equations (A) and (B), into equation (C-3), leads to:

$$u=\frac{1}{2} \tag{C-4}$$

$$v=-\frac{y}{2x} \tag{C-5}$$

As $z=f(x,y)$ varies along the circle, v takes all real values, while $u=\frac{1}{2}$ remains constant, then, the given circle maps onto a vertical line $u=\frac{1}{2}$, as shown in Fig. (**21**).

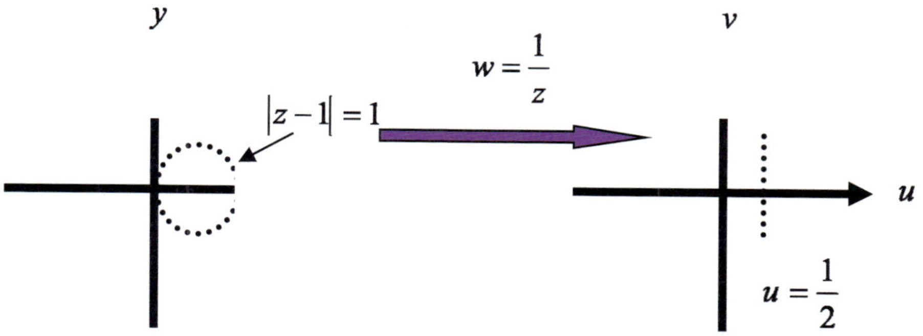

Figure 21: The circle maps to vertical straight line.

Hint

The region inside the circle $|z-1| = 1$ including the circumference if the circle under the transformation $w = f(z) = \frac{1}{z}$ maps to entire region just right to vertical straight line $u = \frac{1}{2}$, as shown in Fig. (**22**).

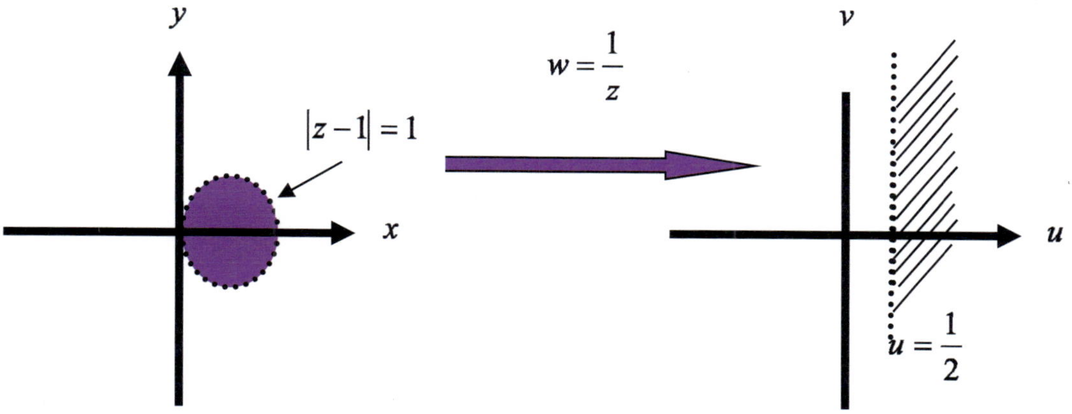

Figure 22: The transformation $w = f(z) = 1/z$ maps to entire region just right to vertical straight line.

Case (3): The Circle $|z| = 2$

The equation of the given circle can be re-written in the following form:

$$x^2 + y^2 = 4 \tag{A}$$

The equations for both u and v, respectively given previously in the following form:

$$u = \frac{x}{x^2 + y^2} \tag{B}$$

$$v = -\left(\frac{y}{x^2 + y^2}\right) \tag{C}$$

Substituting from equation (A) into equation (B), leads to:

$$u = \frac{x}{4}$$

Squaring both sides leads to:

$$u^2 = \frac{x^2}{16} \tag{D}$$

Substituting again from equation (A) into equation (C), leads to:

$$v = -\frac{y}{4}$$

Squaring both sides leads to:

$$v^2 = \frac{y^2}{16} \tag{E}$$

Adding equations (D) and (E), we get:

$$u^2 + v^2 = \frac{1}{16}\left(x^2 + y^2\right) \tag{F}$$

Substituting from equation (A) into equation (F), leads to:

$$u^2 + v^2 = \frac{1}{4} \tag{G}$$

Equation (F) can be re-written in the following form:

$$|w| = \frac{1}{2} \tag{H}$$

Both the original and transformed functions are shown in Fig. (**23**).

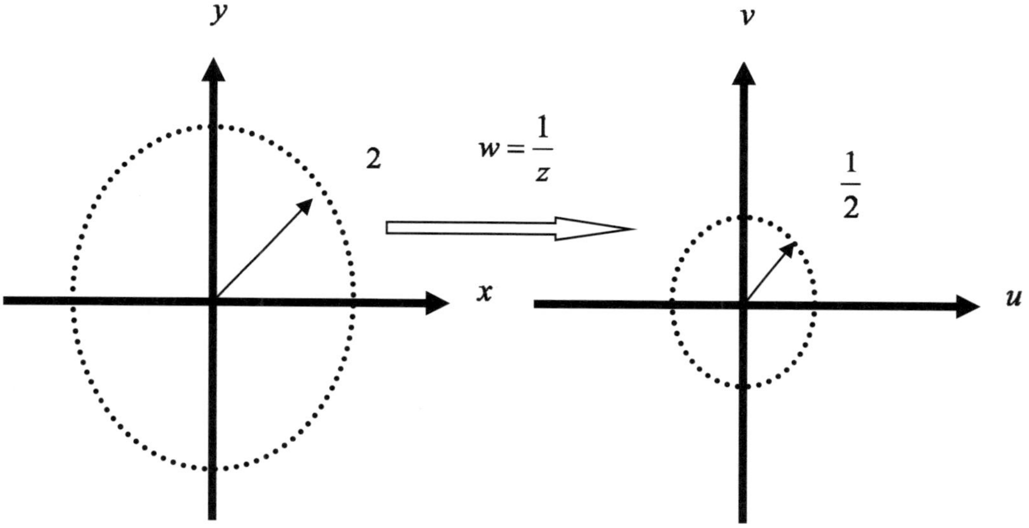

Figure 23: The circle maps to circle but with different radii.

6.5. THE BILINEAR TRANSFORMATION

If a, b, c **and** d are the complex constants, then the function:

$$w = \frac{az + b}{cz + d}, \quad \textbf{such that } ad - bc \neq 0 \tag{12}$$

Equation (12) is called the bilinear transformation. If $c = 0$ then the bilinear transformation reduces to linear transformation. The bilinear transformation possesses a number of very interesting mapping properties. In what follows, we state these properties sequentially **[38]**.

Property Number (1)

Under the bilinear transformation, lines or circles maps lie onto lines or circles. In what follows, we give an outline of this assertion. This outline is based on the following two facts:

1 The bilinear transformation is the composite of the following three functions taken in order:

1.1. The function $\xi = cz + d$, is linear mapping.

1.2. The function $\eta = \dfrac{1}{z}$, is inverse mapping.

1.3. The function $w = \dfrac{a}{c} + \left(\dfrac{bc - ad}{c}\right)\eta$, is inverse mapping.

Fig. (**24**), shows diagrammatically the different stages of bilinear transformation.

Example

Show that the bilinear map $w = \left(\dfrac{z-1}{z+1}\right)$ transforms the half-plane $R(z) \geq 0$ onto unit disk $|w| < 1$.

Hint

Before solution, let us say that this example is of particular interest to scientists in the field of circuit analysis and stability of linear systems.

Solution

Let us first determine the values of the constants a, b, c and d by comparing the given transformation function and the original bilinear transformation as follows:

$$w = \frac{az + b}{cz + d} \Leftrightarrow \frac{z-1}{z+1} \Rightarrow$$

$$a = 1$$

$$b = -1$$

$$c = 1$$

$$d = 1$$

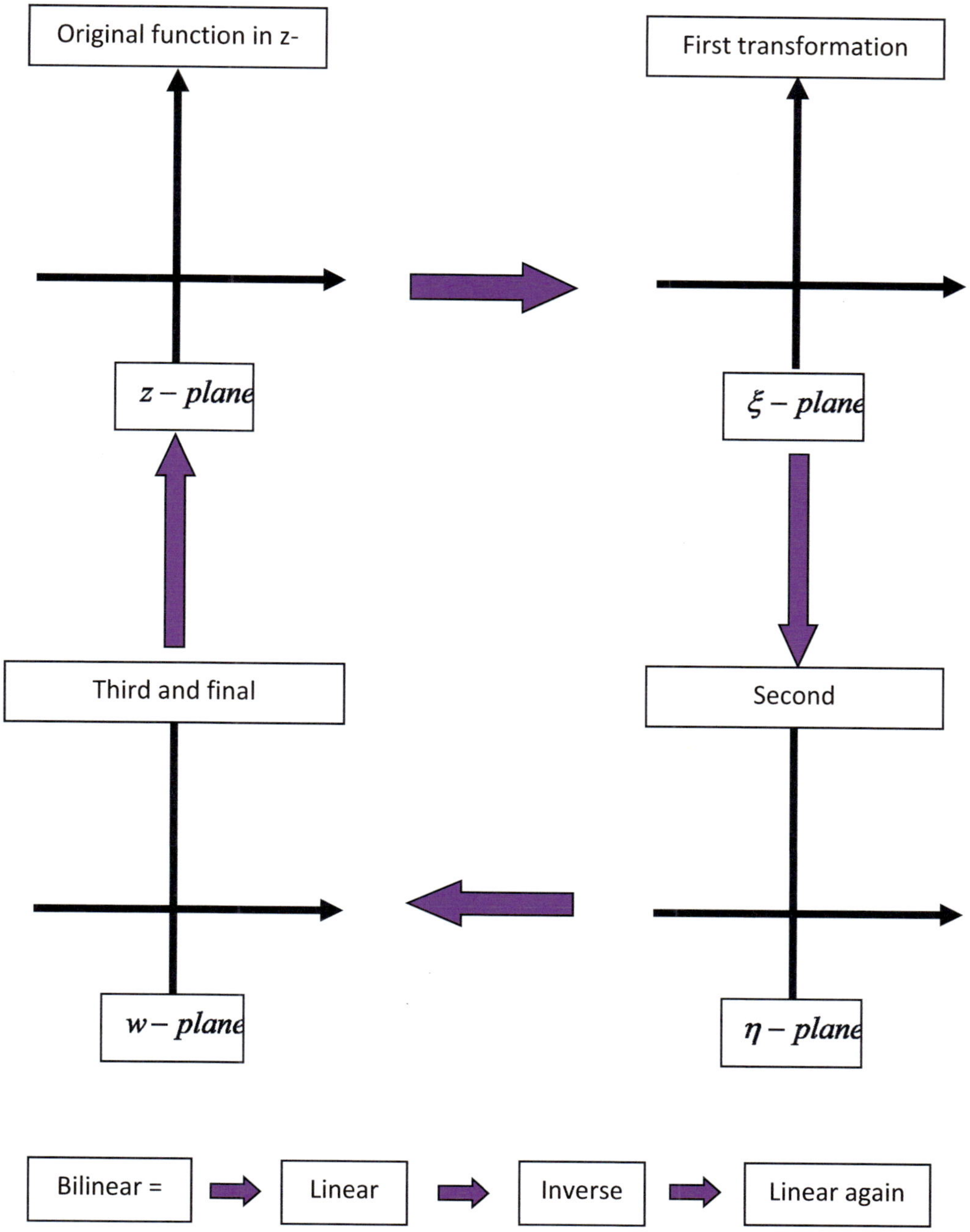

Figure 24: Different stages of bilinear transformation.

The transformation under hand consists of three main steps of transformation, they are:

1 Linear transformation $\xi = cz + d = z + 1$, and this step contains three processes, they are:

1.1. Rotation by $\arg(c), c = 1$

1.2. Magnification by $|c|$

1.3. Translation by $d, d = 1$

Now under this linear transformation, we can conclude the following:

$$\arg(c) = \tan^{-1}\left(\frac{0}{1}\right) = 0 \Leftrightarrow \text{No rotation}$$

$$|c| = 1 \Leftrightarrow \text{No magnification}$$

As long as the term d exists and has a value, this means that there will be shift to the right as shown in Fig. (**25**). This means that under this linear transformation $R(z) \geq 0$ transforms to $R(\xi) > 1$.

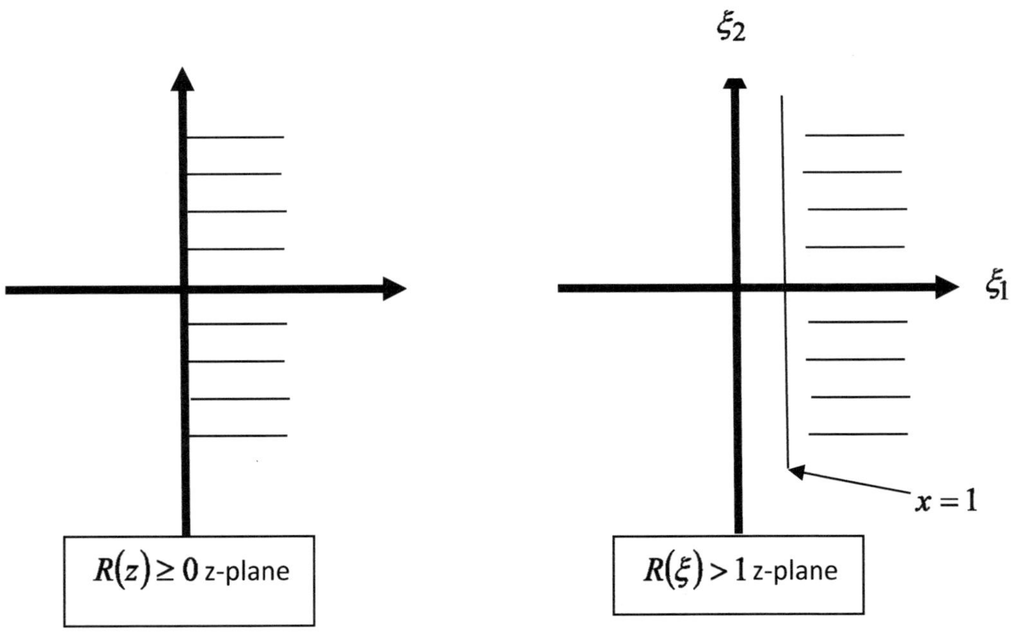

Figure 25: The region in z-plane and the corresponding in ξ -plane after linear transformation.

2. Inverse transformation $\eta = \dfrac{1}{\xi}$

In this part of transformation, we will use the inverse transformation to transform the transformed region obtained from step one, *i.e.*, $R(\xi) \geq 1$. Let us now start the inverse transform procedure:

$$\eta = \frac{1}{\xi} = \frac{1}{\xi_1 + i\xi_2}$$

$$\eta = \frac{1}{\xi_1 + i\xi_2} \times \frac{\xi_1 - i\xi_2}{\xi_1 - i\xi_2} = \frac{\xi_1}{\xi_1^2 + \xi_2^2} - i\frac{\xi_2}{\xi_1^2 + \xi_2^2}$$

$$\eta = \frac{\xi_1}{\xi_1^2 + \xi_2^2} - i\frac{\xi_2}{\xi_1^2 + \xi_2^2}$$

Therefore

$$\eta_1 = \left(\frac{\xi_1}{\xi_1^2 + \xi_2^2}\right)$$

$$\eta_2 = -\left(\frac{\xi_2}{\xi_1^2 + \xi_2^2}\right)$$

Note that:

$$\eta_1 = \left(\frac{\xi_1}{\xi_1^2 + \xi_2^2}\right) = R(\xi)$$

$$\eta_2 = -\left(\frac{\xi_2}{\xi_1^2 + \xi_2^2}\right) = I(\xi)$$

Now, let us dealing with the region in ξ-plane as follows:

$$R(\xi) \geq 1 \Leftrightarrow \xi_1 \geq 1$$

This implies the following:

$$\eta_1 = \left(\frac{1}{1+\xi_2^2}\right) \tag{A}$$

$$\eta_2 = -\left(\frac{\xi_2}{1+\xi_2^2}\right) \tag{B}$$

As we knew before when dealing with inverse transformation that we tried to find a relation connecting the two axes in the transformed plane. Herein we tried to find a relation between η_1 and η_2 by making use of equations (A) and (B). This procedure will be carried out as follows:

From equations (A) and (B), one can get the following relation:

$$\xi_2 = -\eta_1\sqrt{\left(\frac{1}{\eta_1}-1\right)} \tag{C}$$

Squaring both sides of equation (C), leads to:

$$\left(\eta_1 - \frac{1}{2}\right)^2 + \xi_2^2 = \frac{1}{4} \tag{D}$$

Equation (D) is an equation of a circle with center at $\left(\frac{1}{2},0\right)$ and of radius $r = \frac{1}{2}$, as shown in Fig. (**26**). This means that under inverse transformation, the function $R(\xi) \geq 1$ is mapped to circle with center at $\left(\frac{1}{2},0\right)$ and of radius $r = \frac{1}{2}$.

3. Linear Transformation $w = \frac{a}{c} + \frac{bc-ad}{c}\eta$

With the constants $a = 1, b = -1, c = 1, d = 1$, the last linear transformation takes the following form:

$$w = -2\eta + 1 \qquad \text{(A)}$$

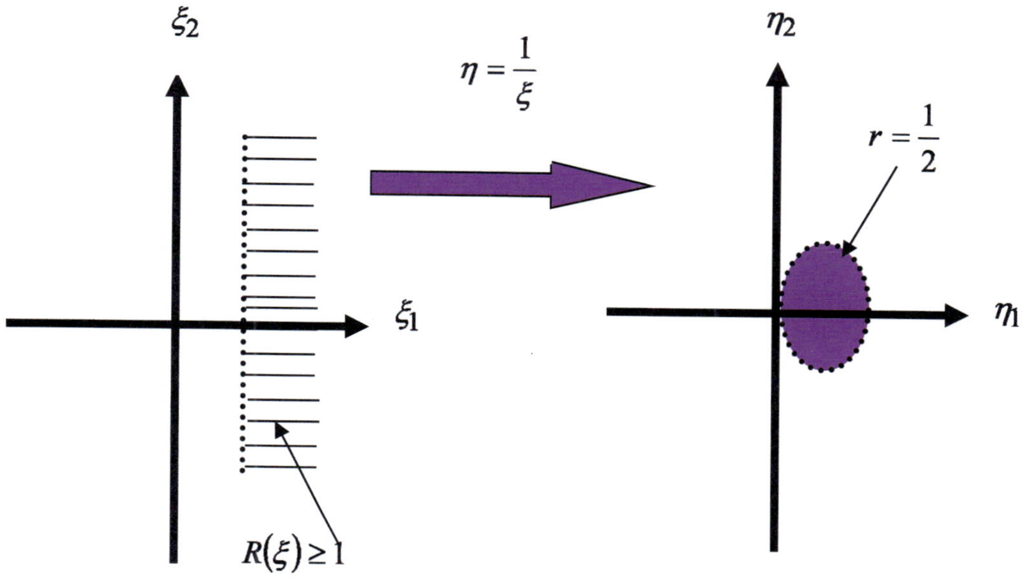

Figure 26: Inverse transformation.

As we said so many times, the linear transformation consists of three main steps; they were according the numerical values of the constants herein:

3.1. Rotation by $\arg(-2)$

3.2. Magnification by $|-2|$

3.3. Translation by 1

Now, let us execute these steps:

3.1. Rotation by $\arg(-2)$

$$\arg(-2) = \tan^{-1}\left(\frac{I}{R}\right) = \tan^{-1}\left(\frac{0}{-2}\right) = -\pi \qquad \text{(B)}$$

The region in η – plane after rotation is shown in Fig. (**27**)

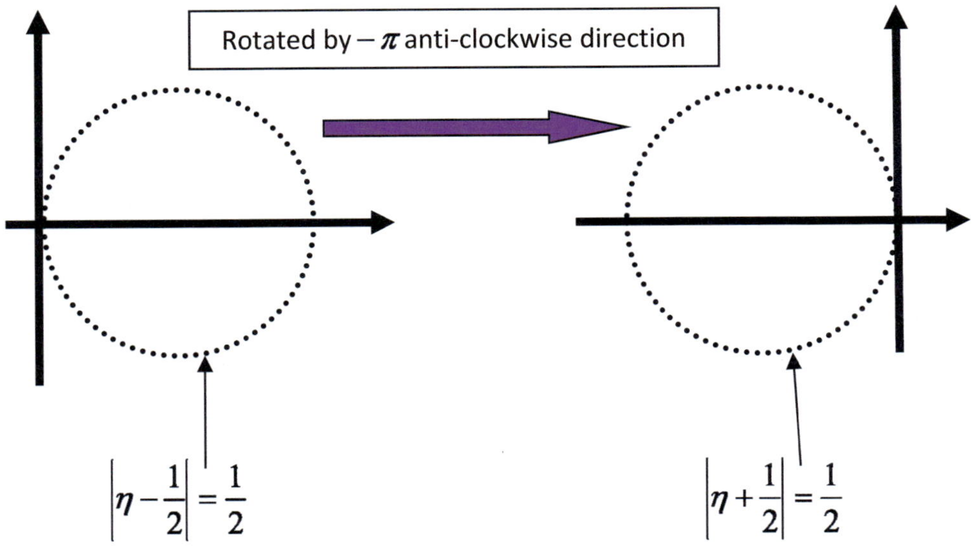

Figure 27: The circle in η -plane before and after rotation.

3.2. Magnification by $|-2|$

Magnification $|-2| = 2$

The region in η – plane after magnification is shown in Fig. (**28**).

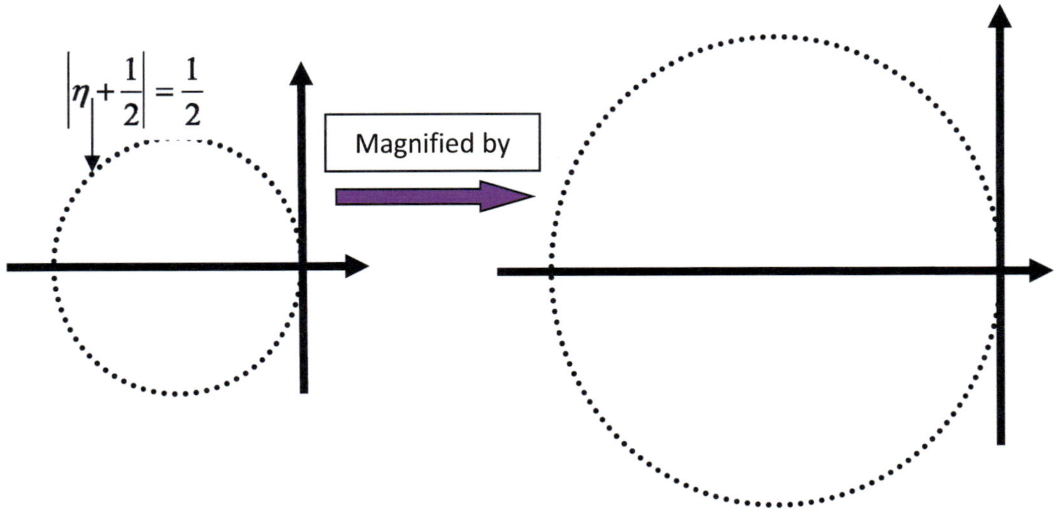

Figure 28: The circle in η -plane before and after magnification.

3.3. Translation by 1

The region after rotation and magnification it will be translated by one unit to the right. The region in η −plane after stretching is shown in Fig. (**29**).

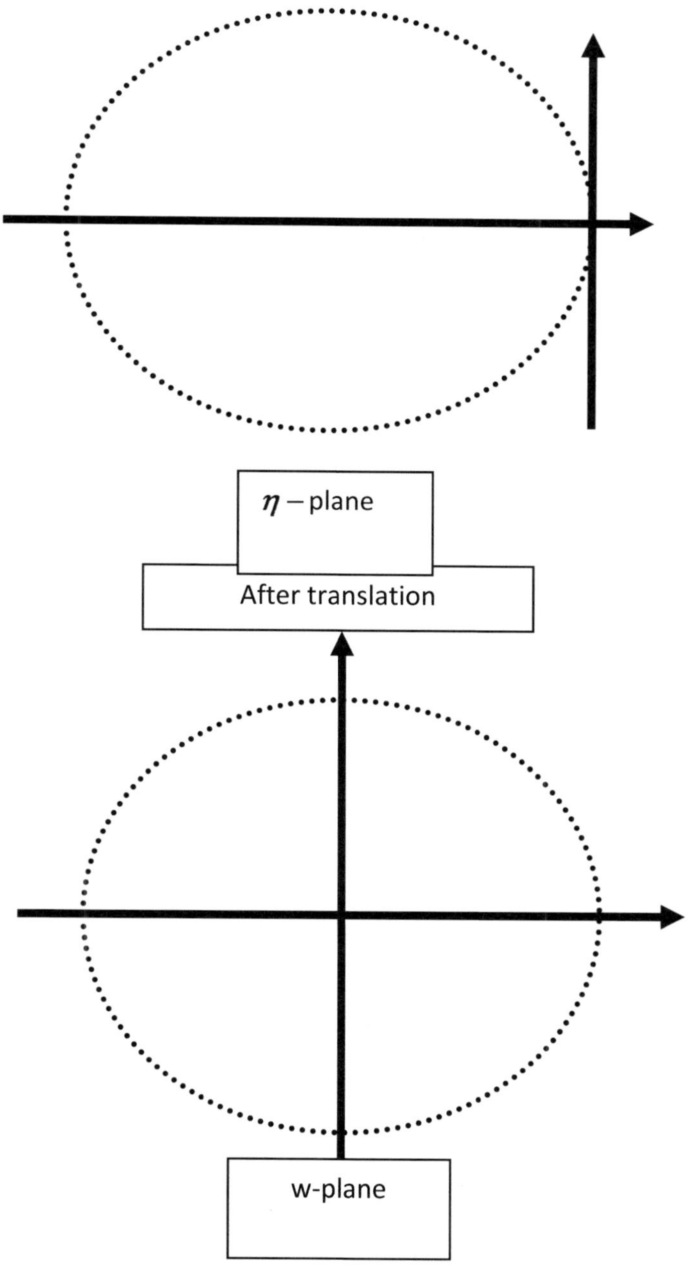

Figure 29: The circle in η -plane and the final in w-plane.

Example

Show that $w = \left(\dfrac{1+z}{-z+1}\right)^{\frac{2}{3}}$ maps a unit circle in the z-plane onto a wedge shape in w-plane.

Solution

First of all, let us make some simplification to the given transformation function so as we can deal with.

$$w = \left(\frac{1+z}{-z+1}\right) = -\left(\frac{1+z}{z-1}\right) = -\left(\frac{z+1}{z-1}\right) = \frac{-z\underbrace{-1+2}-2}{z-1}$$

$$w = \frac{-z\underbrace{-1+2}-2}{z-1} = \frac{(-z+1)-2}{z-1}$$

$$w = \left\{\frac{-(z-1)}{z-1} - \frac{2}{z-1}\right\}$$

$$w = \left\{-1 - \frac{2}{z-1}\right\} \qquad \text{(A-1)}$$

Equation (A-1) is a bilinear transformation, and again it consists of three major transformations, linear, inverse, then linear again. They are:

1. Linear $\Phi = z - 1$
2. Inverse $\Psi = \dfrac{-2}{\Phi}$
3. Linear again $w = -1 + \Psi$

In addition to the three transformations above, there is another transformation corresponding to the power $\left(\dfrac{2}{3}\right)$, *i.e.*,

$$w_1 = w^{\frac{2}{3}}$$

Step (1): Linear Transformation $\Phi = z - 1$

Let us start dealing with the linear transformation $\Phi = z - 1$ and the equation of the circle as follows:

$$\Phi = z - 1$$

$$\Phi = \Phi_1 + i\Phi_2$$

$$z = x + +iy$$

Now then, the equation of the linear transformation takes the form:

$$\Phi_1 + i\Phi_2 = (x-1) + iy$$

This yields to:

$$\Phi_1 = (x-1) \tag{A-2}$$

$$\Phi_2 = y \tag{A-3}$$

From equation (A-2), one can get:

$$x = \Phi_1 + 1 \tag{A-4}$$

And from equation (A-3):

$$y = \Phi_2 \tag{A-5}$$

Equation of the circle in the z-plane is:

$$x^2 + y^2 = 1 \tag{A-6}$$

Now substituting from equations (A-4) and (A-5) into equation (A-6), leads to:

$$(\Phi_1 + 1)^2 + \Phi_2^2 = 1 \tag{A-7}$$

Now, let us perform the three steps of linear transformation, as follows:

$$\arg(a) = \tan^{-1}\left(\frac{0}{1}\right) = 0 \Rightarrow \text{No rotation} \tag{A-8}$$

$$|a| = 1 \Rightarrow \text{No magnification} \tag{A-9}$$

$$b = -1 \Rightarrow \text{There is translation to the left} \tag{A-10}$$

Making use of equations (A-8), (A-9), (A-10) and, under linear transformation $\Phi = z - 1$, the given unit circle in z-plane transformed onto a unit circle of the same radius but with center at $(-1,0)$, as shown in Fig. (**30**).

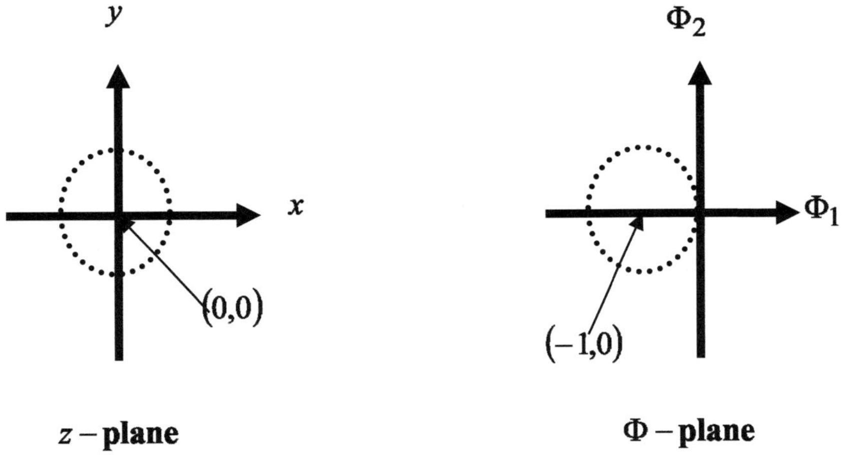

Figure 30: Linear transformation.

Step (2): Inverse Transformation $\Psi = \dfrac{-2}{\Phi}$

Now, we have circle:

$$(\Phi_1 + 1)^2 + \Phi_2^2 = 1 \tag{B-1}$$

We want to transform this circle under the given transformation, to do so, let us fellow up the following:

$$\Psi = \Psi_1 + i\Psi_1 = -2\left(\frac{1}{\Phi_1 + i\Phi_2}\right) \tag{B-2}$$

$$\Psi = -2\left(\frac{1}{\Phi_1 + i\Phi_2} \times \frac{\Phi_1 - i\Phi_2}{\Phi_1 - i\Phi_2}\right) = -2\left(\frac{\Phi_1 - i\Phi_2}{\Phi_1^2 + \Phi_2^2}\right)$$

$$\Psi = \left(\frac{-2\Phi_1}{\Phi_1^2 + \Phi_2^2} + i\frac{2\Phi_2}{\Phi_1^2 + \Phi_2^2}\right)$$

$$\Psi_1 = \frac{-2\Phi_1}{\Phi_1^2 + \Phi_2^2} \tag{B-3}$$

$$\Psi_2 = \frac{2\Phi_2}{\Phi_1^2 + \Phi_2^2} \tag{B-4}$$

Equation (A-7) can now be-written as follows:

$$\Phi_1{}^2 + \Phi_2^2 = -2\Phi_1 \tag{B-5}$$

Divide equation (B-4) by equation (B-3), leads to:

$$\Psi_2 = -\frac{\Phi_2}{\Phi_1} \tag{B-6}$$

But from equation (B-5), one can get:

$$1 = \frac{-2\Phi_1}{\Phi_1{}^2 + \Phi_2^2} \tag{B-7}$$

Now making use of equation (B-7) into the right-hand side of equation (B-3), leads to:

$$\Psi_1 = 1 \tag{B-8}$$

From equations (B-6) and (B-8), we are ready to map the circle in Φ -plane to the transformed Ψ -plane, as shown in Fig. (**31**).

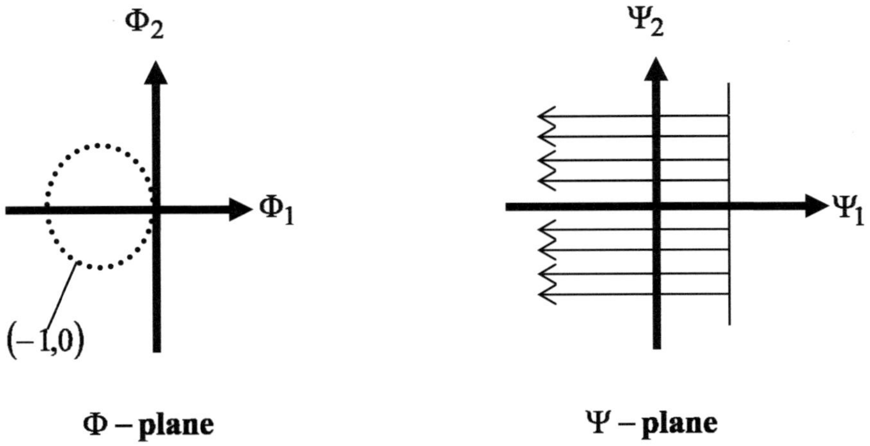

Figure 31: The circle in Φ -plane and the corresponding region in Ψ -plane using inverse transformation.

Step (3): Linear Transformation Again $w=\Psi-1$

As long as it is linear, so we should first determine the constants in the linear transformation, they are:

$$a=1$$
$$b=-1 \qquad \text{(C-1)}$$

We also know, there are three operations in the linear transformation; they are rotation, magnification and finally translation, respectively according to the numerical values of the constants. Now, we have in Ψ -plane the line $\Psi_1=1$ then let us check the existence of any operation corresponds to the linear transformation:

$$\arg(a)=\arg(1)=\tan^{-1}\left(\frac{0}{1}\right)=0 \Leftrightarrow \text{No rotation} \qquad \text{(C-2)}$$

$$|a|=|1|=1 \Leftrightarrow \text{No Magnification} \qquad \text{(C-3)}$$

Let us also see the region in the $w-$**plane** by the following procedure:

$$w = u + iv = (\Psi_1 - 1) + i\Psi_2 \quad \text{(C-4)}$$

$$u = (\Psi_1 - 1) \quad \text{(C-5)}$$

$$v = \Psi_2 \quad \text{(C-6)}$$

But

$$\Psi_1 = 1 \Rightarrow u = 0 \quad \text{(C-7)}$$

$$v = \Psi_2 \quad \text{(C-8)}$$

As long as, there is a value for the constant b this means that there will be a translation equals to -1, *i.e.*, translate the whole region to the left by one unit, as shown in Fig. (**32**).

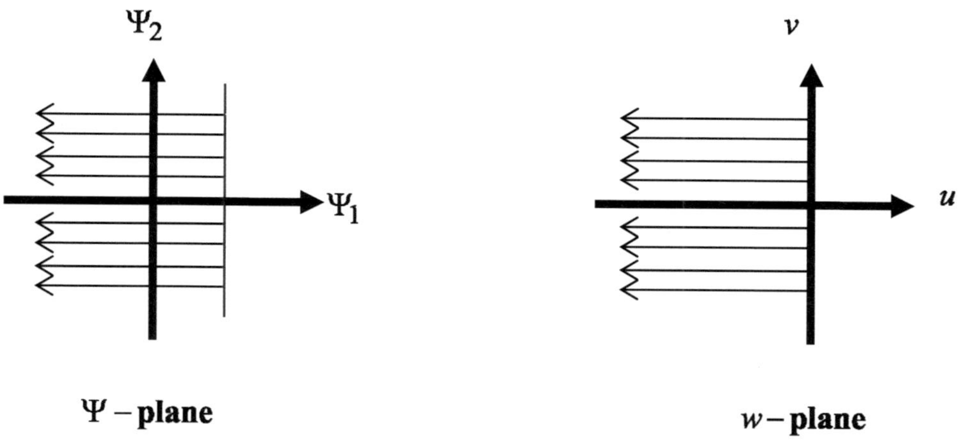

Figure 32: The region in Ψ -plane and the corresponding in w-plane.

Step (4): Power Transformation Again $w_1 = w^{\frac{2}{3}}$

Let us some mathematical manipulation to the given power transformation as follows:

$$w_1 = w^{\frac{2}{3}} \quad \text{(D-1)}$$

$$\underbrace{\arg(w_1)}_{\Theta_1} = \frac{2}{3} \times \underbrace{\arg(w)}_{\Theta} \qquad \text{(D-2)}$$

Note that $\arg(w)$ from $w-$plane was $\frac{\pi}{2}$, then:

$$\arg(w_1) = \frac{2}{3} \times \frac{\pi}{2} = \frac{\pi}{3} \qquad \text{(D-3)}$$

The final region after the four transformation steps is shown in Fig. (**33**).

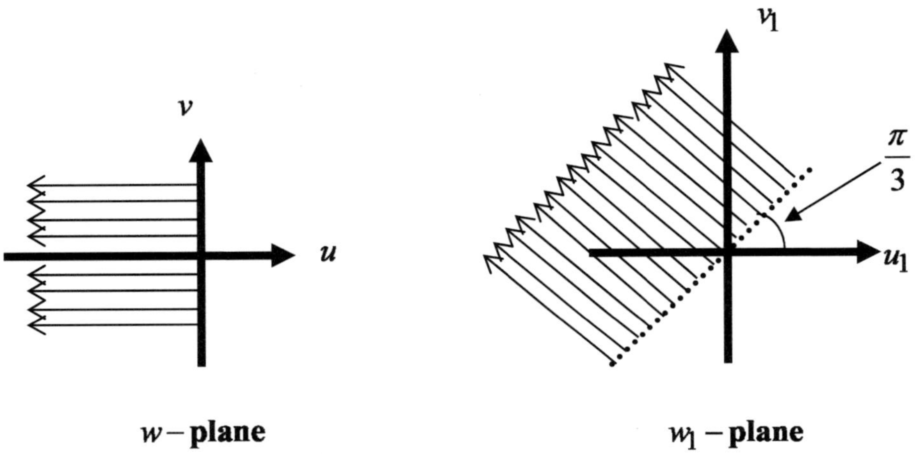

Figure 33: The final region after the four transformation steps.

This means that the initial unit circle in z-plane after previous four transformation steps transformed to wedge.

Property Number (2)

The second property of the bilinear transformation can be described as follows:

"Given any triple z_1, z_2, z_3 ***of distinct points in z-plane, and any triple*** w_1, w_2, w_3 ***of distinct points in the w-plane, there is a unique bilinear transformation mapping*** z_j ***onto*** w_j, $j = 1,2,3$ ***"***

In order to find the mapping whose existence is asserted above, we use the cross-ration given by the following form:

$$\frac{(w-w_1)(w_2-w_3)}{(w-w_3)(w_2-w_1)}=\frac{(z-z_1)(z_2-z_3)}{(z-z_3)(z_2-z_1)} \tag{13}$$

To remember the above equality, see Fig. (**34**):

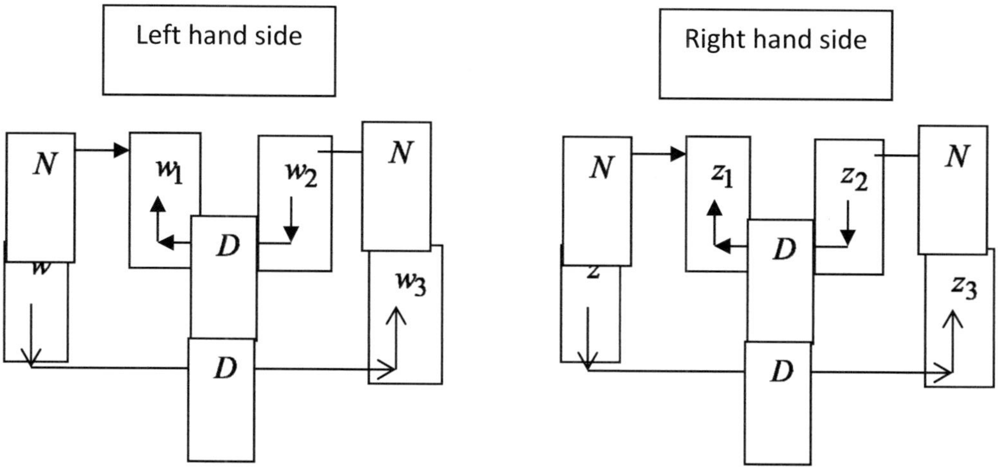

Figure 34: Empirical method to construct cross-ratio given by (13).

Some notations in Fig. (**34**)

N: Numerator

D: Denominator

Example

Find the bilinear transformation that will map the points $z_1 = 0, z_2 = i, z_3 = -1$ onto $w_1 = 12, w_2 = 11+i, w_3 = 11$, respectively.

Solution

Let us first, start the solution by write down the cross-ratio relation:

$$\frac{(w-w_1)(w_2-w_3)}{(w-w_3)(w_2-w_1)}=\frac{(z-z_1)(z_2-z_3)}{(z-z_3)(z_2-z_1)} \tag{A}$$

Substituting the numerical values given, leads to:

$$\frac{(w-12)i}{(w-11)(-1+i)} = \frac{z(1+i)}{(z+1)i} \qquad \text{(B)}$$

After simplifying equation (B), one will get the following form:

$$w = \frac{10z-12}{z-1} \qquad \text{(C)}$$

Exercises

1. State where the following mappings are conformal:

(a) $w = \exp(z)$

(b) $w = \sin z$

(c) ..

(d) $w = \exp(z^2+1)$

(e) $w = \dfrac{1}{z}$

(f) $w = \dfrac{z+1}{z-1}$

For the problems (2-5), find the angle of rotation and the scale factor of the mapping $w = f(z)$ at the indicated points:

2. $w = \dfrac{1}{z}$ at the points $1, 1+i$ and i

3. $w = \ln r + i\Theta$, where $\dfrac{-\pi}{2} < \Theta < \dfrac{3\pi}{2}$ at the points $1, 1+i$ and -1

4. $w = r^{\frac{1}{2}} \cos\left(\dfrac{\Theta}{2}\right) + i r^{\frac{1}{2}} \sin\left(\dfrac{\Theta}{2}\right)$, where $-\pi < \Theta < \pi$ at the points $i, 1, -i$ and $3+4i$

5. $w = \sin z$, at the points $\frac{\pi}{2} + i$, 0 and $\frac{-\pi}{2} + i$

6. Consider the mapping $w = z^2$. If $a \neq 0$ and $b \neq 0$, show that the lines $x = a$ and $y = b$ are mapped onto orthogonal parabolas.

7. Consider the mapping $w = z^{\frac{1}{2}}$, where $z^{\frac{1}{2}}$ denotes the principle branch of the square root function. If $a > 0$ and $b > 0$, show that the lines $x = a$ and $y = b$ are mapped onto orthogonal curves.

8. Consider the mapping $w = \exp(z)$. Show that the lines $x = a$ and $y = b$ are mapped onto orthogonal curves.

9. Show that $w = S(z) = \frac{i(1-z)}{(1+z)}$ maps the unit disk $|z| < 1$ one-to-one and onto the upper half-plane $\mathrm{Im}(w) > 0$.

10. Construct the bi-linear transformation $w = S(z)$ that maps the points $z_1 = -i, z_2 = 1$ and $z_3 = i$ onto the points $w_1 = -1, w_2 = 0$ and $w_3 = 1$, respectively.

11. Find the bi-linear transformation $w = S(z)$ that maps the points $z_1 = -2$, $z_2 = -1-i$ and $z_3 = 0$ onto the points $w_1 = -1$, $w_2 = 0$ and $w_3 = 1$, respectively.

12. Show that the mapping $w = S(z) = \frac{(1-i)z+2}{(1+i)z+2}$ maps the disk $D: |z+1| < 1$ onto the upper half-plane $\mathrm{Im}(w) > 0$.

13. Find the bi-linear transformation that maps the crescent-shaped region that lies inside the disk $D: |z-2| < 2$ and outside the circle $|z-1| = 1$ onto a horizontal strip.

REFERENCES

REFERENCES

[1] R. Remmert, *Theory of Complex Functions.* Springer, 1991.
[2] J.H. Mathews and R.W. Howel, *Complex Analysis for Mathematics and Engineering.* 5th ed, Jones and Bartlett Publishers, 2006.
[3] R. Remmert, *Classical Topics in Complex Function Theory.* Springer, 2010.
[4] R.P. Boas, *Invitation to Complex Analysis, Hardbound.* 2010 Series: MAA Textbooks, 2010.
[5] T. Gamelin, *Complex Analysis.* August 19, 2003.
[6] R. Narasimhan and Y. Nievergelt, *Complex Analysis in One Variable.* December 21, 2000.
[7] W. Rudin, *Real Complex Analysis.* Published May 1st 1986 by McGraw-Hill Science/Engineering/Math (first published January 1st 1970).
[8] W.T. Shaw, *Complex Analysis with Mathematica.* Cambridge University Press, June 2006.
[9] L. Ahlfors, *Complex Analysis.* 3rd ed., McGraw-Hill, 1979.
[10] C. Carathéodory, *Theory of Functions of a Complex Variable.* 2 volumes, Chelsea: New York.
[11] T. Needham, *Visual Complex Analysis.* Oxford, 1997.
[12] P. Henrici, *Applied and Computational Complex Analysis* (Wiley). 3-volumes, 1974, 1977, 1986.
[13] E. Kreyszig, *Advanced Engineering Mathematics.* 9 ed., Ch.13-18 Wiley, 2006.
[14] A.I. Markushevich, *Theory of Functions of a Complex Variable.* Prentice-Hall, 1965, 3 volumes.
[15] V. Scheidemann, *Introduction to Complex Analysis in Several Variables.* Birkhauser, 2005.
[16] M.R. Spiegel, *Theory and Problems of Complex Variables - with an introduction to Conformal Mapping and its applications.* McGraw-Hill, 1964.
[17] Marsden & Hoffman, *Basic Complex Analysis. Freeman,* 1999.
[18] J.W. Brown and R.V. Churchill, *Complex Variables and Applications.* 8th ed. McGraw-Hill: 2008.
[19] R. Remmert, *Theory of Complex Functions.* ISBN: 9780387971957, Springer-Verlag
[20] D.G. Zill and P.D. Shanahan, *A First Course in Complex Analysis with Applications.* Jones and Bartlett Publishers Series in Mathematics, 25 Feb 2009.
[21] T. Gamelin, *Complex Analysis.* Springer, 2001
[22] R. Greene and S. Krantz, *Function Theory of One Complex Variable.* 3rd ed., AMS, 2006.
[23] E. Hille, *Analytic Function Theory.* vols. I and II, Chelsea, 1973
[24] R. Narasimhan and Y. Nievergelt, *Complex Analysis in One Variable.* 2nd ed., Birkhauser, 2001
[25] W. Rudin, *Real and Complex Analysis.* 3rd ed., McGraw-Hill, 1987.
[26] E. Stein and R. Shakarchi, *Complex Analysis.* Princeton University Press, 2003.
[27] O. Forster, *Lectures on Riemann Surfaces.* Springer, 1981.
[28] S. Krantz, *Complex Analysis: The Geometric Viewpoint.* Carus Math Monographs no. 23, MAA, 1990.
[29] Z. Nehari, *Conformal Mappings.* Dover, 1990.
[30] R. Remmert, *Classical Topics in Complex Function Theory.* Springer, 1998.
[31] R. Kühnau, *Handbook of Complex Analysis.* Volume 2, Pages 1-861, 2005.
[32] T. Gamelin, *Complex Analysis.* Springer, 2003.
[33] T. Needham, *Visual Complex Analysis.* Oxford University Press, 1999.
[34] E.M. Stein and R. Shakarchi, *Complex Analysis.* Princeton University Press, 2003.
[35] Ablowitz and Fokas, *Complex Variables.* 1997
[36] Carrier, Krook and Pearson, *Functions of a Complex Variable.* McGraw-Hill, 1966.
[37] Stewart and Tall, *Complex Analysis.* Cambridge, 1996.
[38] Anant R. Shastri, *Basic Complex Analysis of One Variable.* Jan 2010.

S. G. Ahmed

REFERENCES

Subject Index

A

Algebra of Complex Numbers 5
Analytical functions 29, 49, 65-66, 103, 117, 136, 149
Annular region 66
Annulus theorem 61, 66, 68
Application of residue theory 117, 122

B

Bilinear transformation 29, 149, 175-77, 184, 190-91

C

Cauchy-Riemann equations 48-49, 51
Cauchy theorem 29, 65, 76, 128-29, 131, 136
Cauchy theorem of integration 76, 129, 131, 136
Circles maps 176
Circular arc 154, 159-61
Closed annulus region 66
Closed interval 86-87
Complex exponential function 38-40, 53
Complex function 29-30, 48-49, 82, 86, 90, 92, 94, 103, 107, 110-11, 113, 126, 132, 134
Complex function-definition 29
Complex functions 108
Complex integration 61, 97, 117, 122
Complex-valued function 29, 38, 56
Complex z-plane 24
Constant of integration 51
Continuous functions 84, 94
Corresponding complex functions 131
Cosh function 54
Cubic equation 1

D

Dashed region 117-18
Depressed equation 1
Differentiable complex function 86
Domain problem 67-68, 77

E

Elementary functions 38
Expansion of functions 105
Exponential function 40, 42, 54

F

Finite number of poles 126, 131

G

Generalized Cauchy 61, 73-76

H

Harmonic complex function 49
Harmonic functions 51
Holder inequalities 16
Horizontal axes 20, 126, 157, 160

I

Imaginary parts 9, 16, 47, 86-87, 134, 141, 163
Improper integrals 117, 126, 131-32, 134
Initial unit circle 190
Integer values 22-23
Integral formula 73-76
Integrals of rational functions of sine 122
Integral theorem 61, 77-78
Interior circle 24, 26
Interior path 68, 70
Inverse mapping 33, 36, 176
Inverse transformation 29, 35, 149, 169-70, 179-81, 186, 188
Isolated singularity 110-11, 117-18

L

Laurent complex series 103
Laurent expansion 110, 117-18
Laurent series 107-11, 114, 117
Linear transformation 29, 36-38, 59, 149, 152-53, 175, 178, 180-81, 185-86, 188
Line integral 61-63, 65, 94
Line segment 90, 126, 154, 157-59, 166-67
old 157-58
Logarithmic complex functions 29, 42
Logarithmic functions 53-54

M

Maclaurin 103, 105
Maclaurin complex series 103
Maclaurin expansions 105
Maclaurin series 105-6, 110, 113
Magnification rotation magnification 156
Mathematical manipulation 32, 55, 57, 78, 170, 189
Modulus 9, 11, 162
Morera's theorem 70
Multiplication process 5
Multi-valued function log 42, 45

S. G. Ahmed

N

Natural logarithms 53-54
Negative real axis 44
Non-removable isolated singularity 118
Non-zero complex numbers 41-43, 45

O

One-to-One Function 32
One-to-one mapping 33, 35-36, 151
Order of order 126, 131
Oriented circle 119, 124
Original angle 157, 160
Original bilinear transformation 176
Original circular arc 159-60
Original line segment 157
Orthogonal curve 193

P

Parametric form 84, 86-88, 90, 92
Parametric functions 90
Partial fractions 78, 96-97, 108
Polar form 20-22, 59, 162
Pole of order 111-12, 119
Poles existence 128
Positive closed path 73
Positive orientation 67-68, 95-97, 99-100
Power transformation 149, 162, 189
Problem configuration 63-64

Q

Quadratic equation 53-54

R

Rational function 121-22, 132
Real definite integrals 117, 122
Real domain 61
Real exponential function 38
Real integrals 117, 122
Real poles 127-28
Real-valued functions 82, 117
Real variables 38, 82, 122
Region of integration 77-78
Removable singularity 111
Representation of complex number 21
Residue theorem 117-18
Riemann mapping theorem 29
Roots of complex number 22
Rotated magnification 156, 158
Rotated magnification translation 156, 159, 161
Rotation magnification 158-59, 161-62
Rotation magnification translation 156, 159, 162
Rotation stretching 152

S

Simple closed path 66, 94
Simple poles 112
Simple trigonometric functions 105
Sine function 53
Single-valued exponential function 39
Single-valued functions 62
Solution of Part 17-18
Spatial variables 49-51
Square root functions 193

T

Taylor complex series 103
Taylor series 103-4, 112, 114
Transformation 29, 31, 33, 41, 149-50, 153, 163, 166, 169-70, 173, 178-79, 184
  bi-linear 193
Transformation functions 176, 184
Transformation steps 190
Transformed functions 175
Translation 33, 152-54, 156, 159-60, 162, 178, 183, 186, 188-89
Trigonometric functions 105, 122

Made in the USA
Monee, IL
08 May 2026

49715006R00121